NATIONAL ACADEMIES

Sciences
Engineering
Medicine

NATIONAL
ACADEMIES
PRESS
Washington, DC

Planetary Protection Considerations for Missions to Solar System Small Bodies

Report Series—Committee on Planetary Protection

Committee on Planetary Protection

Space Studies Board

Division on Engineering and
Physical Sciences

Board on Life Sciences

Division on Earth and Life Studies

Consensus Study Report

NATIONAL ACADEMIES PRESS 500 Fifth Street, NW Washington, DC 20001

This study is based on work supported by Contract NNH17CB02B/NNH17CB01T with the National Aeronautics and Space Administration. Any opinions, findings, conclusions, or recommendations expressed in this publication do not necessarily reflect the views of any agency or organization that provided support for the project.

International Standard Book Number-13: 978-0-309-69372-1
International Standard Book Number-10: 0-309-69372-1
Digital Object Identifier: https://doi.org/10.17226/26714

Copies of this publication are available free of charge from:

Space Studies Board
National Academies of Sciences, Engineering, and Medicine
Keck Center of the National Academies
500 Fifth Street, NW
Washington, DC 20001

This publication is available from the National Academies Press, 500 Fifth Street, NW, Keck 360, Washington, DC 20001; (800) 624-6242 or (202) 334-3313; http://www.nap.edu.

Suggested citation: National Academies of Sciences, Engineering, and Medicine. 2023. *Planetary Protection Considerations for Missions to Small Bodies in the Solar System: Report Series—Committee on Planetary Protection.* Washington, DC: The National Academies Press. https://doi.org/10.17226/26714.

The **National Academy of Sciences** was established in 1863 by an Act of Congress, signed by President Lincoln, as a private, nongovernmental institution to advise the nation on issues related to science and technology. Members are elected by their peers for outstanding contributions to research. Dr. Marcia McNutt is president.

The **National Academy of Engineering** was established in 1964 under the charter of the National Academy of Sciences to bring the practices of engineering to advising the nation. Members are elected by their peers for extraordinary contributions to engineering. Dr. John L. Anderson is president.

The **National Academy of Medicine** (formerly the Institute of Medicine) was established in 1970 under the charter of the National Academy of Sciences to advise the nation on medical and health issues. Members are elected by their peers for distinguished contributions to medicine and health. Dr. Victor J. Dzau is president.

The three Academies work together as the **National Academies of Sciences, Engineering, and Medicine** to provide independent, objective analysis and advice to the nation and conduct other activities to solve complex problems and inform public policy decisions. The National Academies also encourage education and research, recognize outstanding contributions to knowledge, and increase public understanding in matters of science, engineering, and medicine.

Learn more about the National Academies of Sciences, Engineering, and Medicine at **www.nationalacademies.org**.

COMMITTEE ON PLANETARY PROTECTION

JOSEPH K. ALEXANDER, Alexander Space Policy Consultants, *Co-Chair*
AMANDA R. HENDRIX, Planetary Science Institute, *Co-Chair*
ANGEL ABBUD-MADRID, Colorado School of Mines
ANTHONY COLAPRETE, NASA Ames Research Center
MICHAEL J. DALY, Uniformed Services University of the Health Sciences
DAVID P. FIDLER, Council on Foreign Relations
SARAH A. GAVIT, Jet Propulsion Laboratory
ANDREW D. HORCHLER, Astrobotic Technology, Inc.
EUGENE H. LEVY, Rice University
ROBERT E. LINDBERG, JR., Independent Consultant
MARGARITA M. MARINOVA,[1] Project Kuiper
A. DEANNE ROGERS, Stony Brook University, The State University of New York
GERHARD H. SCHWEHM, European Space Agency (retired)
TRISTA J. VICK-MAJORS, Michigan Technological University

Staff

DANIEL NAGASAWA, Program Officer, Space Studies Board, *Study Director*
NANCY CONNELL, Senior Scientist, Board on Life Sciences
ALEXANDER BELLES, Mirzayan Science & Technology Policy Graduate Fellow, Space Studies Board
MEGAN CHAMBERLAIN, Senior Program Assistant, Space Studies Board
COLLEEN N. HARTMAN, Director, Space Studies Board and Aeronautics and Space Engineering
    Board

---

[1] Recused from this study.

Reviewers

This Consensus Study Report was reviewed in draft form by individuals chosen for their diverse perspectives and technical expertise. The purpose of this independent review is to provide candid and critical comments that will assist the National Academies of Sciences, Engineering, and Medicine in making each published report as sound as possible and to ensure that it meets the institutional standards for quality, objectivity, evidence, and responsiveness to the study charge. The review comments and draft manuscript remain confidential to protect the integrity of the deliberative process.

We thank the following individuals for their review of this report:

Linda T. Elkins-Tanton (NAS), Arizona State University,
Victoria E. Hamilton, Southwest Research Institute,
Henry Hsieh, Planetary Science Institute,
G. Scott Hubbard, Stanford University,
Chris Lewicki, Planetary Resources,
Harry "Hap" Y. McSween (NAS), University of Tennessee,
Carol Raymond, Jet Propulsion Laboratory, and
Edgard G. Rivera-Valentin, Lunar and Planetary Institute.

Although the reviewers listed above provided many constructive comments and suggestions, they were not asked to endorse the conclusions or recommendations of this report nor did they see the final draft before its release. The review of this report was overseen by A. Thomas Young (NAE), Lockheed Martin Corporation (retired), and Melissa A. McGrath, SETI Institute. They were responsible for making certain that an independent examination of this report was carried out in accordance with the standards of the National Academies and that all review comments were carefully considered. Responsibility for the final content rests entirely with the authoring committee and the National Academies.

Contents

PREFACE   xiii

EXECUTIVE SUMMARY   1

1   INTRODUCTION   4

2   SMALL BODIES: BACKGROUND AND CONSIDERATIONS   8

3   CRITERIA FOR PLANETARY PROTECTION CATEGORIZATION OF
    SMALL BODY MISSIONS   27

4   IMPLICATIONS OF PLANETARY PROTECTION CATEGORY I VERSUS
    CATEGORY II FOR SMALL BODY MISSIONS   31

5   PLANETARY PROTECTION, SMALL SOLAR SYSTEM BODIES, AND
    COMMERCIAL SPACE ACTIVITIES   34

APPENDIXES

A   Statement of Task   41
B   COSPAR Planetary Protection Requirements for Category I and Category II Missions 43
C   Acronyms and Abbreviations   44
D   Committee and Staff Biographies   46

Preface

The Space Studies Board (SSB; and its predecessor, the Space Science Board) of the National Academies of Sciences, Engineering, and Medicine has been involved in shaping the U.S. planetary protection policy for 60 years. Through those years, the National Aeronautics and Space Administration (NASA) has sponsored studies through the SSB, seeking independent, scientific advice on how to craft its planetary protection policies. NASA's policies, in turn, have formed a basis upon which the global space science community has developed consensus international planetary protection policies through the International Council for Science's Committee on Space Research (COSPAR).

In 2016, NASA asked the SSB to perform a study on the development of planetary protection policies. The resultant report, *Review and Assessment of Planetary Protection Policy Development Processes*, was released in 2018, followed by a separate 2019 report of NASA's Planetary Protection Independent Review Board.[1] Both studies concluded that there was a need for NASA to "reestablish an independent and appropriate advisory body and process to help guide formulation and implementation of planetary protection adequate to serve the best interests of the public, the NASA program, and the variety of new entrants that may become active in deep space operations in the years ahead."[2] At NASA's request, the newest discipline committee of the SSB was formed in July 2020, the Committee on Planetary Protection (CoPP), to serve as the standing forum for the discussion of planetary protection issues critical to NASA.

For the committee's third report, NASA's Science Mission Directorate and Office of Safety and Mission Assurance leadership requested that the CoPP draft a report discussing the planetary protection considerations for outbound missions to small bodies in the solar system. More specifically, NASA asked CoPP to consider whether identifiable populations within the solar system are so numerous or of sufficiently low scientific interest as to merit relief from planetary protection requirements in order to simplify possible future exploration and commercial ventures. Given interest in developing a robust exploration program, as discussed in the Planetary Science and Astrobiology Decadal Survey 2022,[3] and in the burgeoning private space sector, this is a timely issue to address.

To gather information and discuss the issues, the CoPP met five times in 2021 and 2022, virtually during the COVID-19 pandemic, on the following dates: November 30 to 2 December 2021; January 19, 2022; February 9, 2022; March 2, 2022; and March 21–24, 2022. A completed draft of this report was assembled in April 2022.

The committee would like to thank J. Nick Benardini (NASA), Athena Coustenis (COSPAR Panel on Planetary Protection), Lori Glaze (NASA), Stefanie Milam (Small Bodies Assessment Group), Harry McSween (University of Tennessee), Daniel Britt (University of Central Florida), Clark Chapman (Southwest Research Institute [SwRI]), David Trilling (Northern Arizona University), Lori Feaga (University of Maryland), Henry Hsieh (Planetary Science Institute), William Bottke (SwRI), Cathy Olkin (SwRI), Kelvin Coleman (Federal Aviation Administration [FAA]), Jeff Parker (Advanced Space), Joel Sercel (TransAstra), Robert Jedicke (University of Hawaii), Andy Rivkin (Johns Hopkins University), Brent Buffington (Jet Propulsion Laboratory-California Institute of Technology [JPL-Caltech]), Mark Wallace (JPL-Caltech), Jason Dworkin (NASA Goddard Space Flight Center), Alexandra

---

[1] National Academies of Sciences, Engineering, and Medicine (NASEM), 2020, *Assessment of the Report of NASA's Planetary Protection Independent Review Board*, Washington, DC: The National Academies Press, https://doi.org/10.17226/25773.

[2] NASEM, 2018, *Review and Assessment of Planetary Protection Policy Development Processes*, Washington, DC: The National Academies Press, https://doi.org/10.17226/25172, Recommendation 3.6.

[3] NASEM, 2022, *Origins, Worlds, and Life: A Decadal Strategy for Planetary Science and Astrobiology 2023–2032*, Washington, DC: The National Academies Press, https://doi.org/10.17226/26522.

Pontefract (Georgetown University), Harrison Smith (Earth-Life Science Institute), Frank Groen (NASA), Steph Earle (FAA), Tom Hammond (U.S. House of Representatives Subcommittee on Space and Aeronautics), Ezinne Uzo-Okoro (Office of Science and Technology Policy), and Pamela Whitney (U.S. House of Representatives Subcommittee on Space and Aeronautics) for their presentations to CoPP.

Executive Summary

The ultimate goal of planetary protection for outbound missions is to prevent harmful contamination that would inhibit future measurements designed to search for evidence of the existence or evolution of extraterrestrial life. This goal is codified by the Outer Space Treaty of 1967, consensus guidance from international scientific organizations such as the Committee on Space Research (COSPAR), and National Aeronautics and Space Administration (NASA) policy. Preventing harmful contamination is achieved by following specific guidelines based on existing scientific knowledge about the destination and the type of mission. Planetary protection policy categorizes missions according to the intended targets (e.g., Mars) and mission objectives (e.g., flyby, orbiter, or lander).

Small bodies discussed in this report include Main Belt asteroids (MBAs), near-Earth objects (NEOs), Trojan asteroids, comets, Centaurs, and Kuiper Belt objects (KBOs). Exploration of some of these objects has been performed in the past by government space agencies such as NASA to understand the nature of these objects and what they can reveal about the formation of the solar system. These studies, including both ground- and space-based astronomical observations, compose the bulk of knowledge about these objects to this date, and have been conducted from a purely scientific perspective.

With the advent of nongovernmental space missions, including by private-sector enterprises, that paradigm will change. Private entities are preparing to expand space operations to include mining small bodies, especially NEOs, to obtain resources for future space exploration. Therefore, at this critical juncture, it is timely to reassess planetary protection requirements for missions to these types of objects.

NASA and COSPAR currently consider missions to undifferentiated, metamorphosed asteroids as under planetary protection Category I and missions to comets and all other types of asteroids as under Category II. Missions to KBOs are considered Category II by both NASA and COSPAR. Neither COSPAR nor NASA cite a categorization for missions to Centaurs. No requirements for spacecraft cleanliness or organics inventories are imposed on either Category I or Category II missions to small bodies. Category I missions carry no planetary protection requirements at all. Category II only requires the provision of information about the mission important for planetary protection, such as the mission's intended target, the spacecraft's trajectory, and the final deposition of the spacecraft.

The committee learned that there are often misconceptions within the commercial space sector about these requirements. There is the incorrect perception that missions to small bodies are required to have material inventories or to use special cleanroom procedures to meet planetary protection requirements. That is not the case. Such decisions are only motived by either (1) the sensitivity or reliability requirements of the mission itself or (2) a mission trajectory that includes a flyby of a body more sensitive to biological contamination (e.g., Mars) and thus merit a more stringent categorization. Engineering missions have cleanliness requirements driven by spacecraft reliability.

The committee also found that confusion extends into the scientific community with regards to the planetary protection requirements for missions to small bodies. Members of the community were found to believe that planetary protection measures for missions to small bodies were in place to prevent contamination whereas Category II missions do not have any such requirements. Generally, science missions have cleanliness requirements that are not dictated by planetary protection but by the nature of the scientific measurements to be performed.

This report responds to NASA's request for a study on planetary protection categorization of missions to small bodies, including whether there are particular populations of small bodies for which contamination of one object in the population would not be likely to have a tangible effect on the opportunities for scientific investigation using other objects in the population. NASA also asked that, if such populations exist, would it be suitable to categorize future missions to those bodies as Category I.

In addressing NASA's request, the committee considered surface composition of target bodies and their importance for prebiotic chemistry, along with size of the small-body populations, the current state of knowledge on the types of objects, the likelihood of a future scientific mission returning to any specific object, active object surface processes, and the size. Each factor has varying degrees of importance to the consideration of planetary protection categorization.

The committee offers these specific findings related to the statement of task of this study:

Finding 1: The primary astrobiological value of small solar system bodies is that some of these bodies contain prebiotic organic compounds that are relevant to the study of the origin of life in the solar system.

Finding 2: Based on current knowledge, it is highly improbable that small bodies harbor extinct or extant life, or that terrestrial microbes carried by a landing spacecraft can proliferate on a small body. Furthermore, given the short timescales of inactivation by ultraviolet C (UVC, light in the 200–280 nm) radiation, there is no realistic likelihood that terrestrial microbes delivered by a spacecraft to a particular small body can be transported to another small body in a timeframe comparable to the timescales relevant for missions to small bodies (i.e., contaminating body A will not threaten body B).

Finding 3: The committee does not find a need to change current categorization of missions to small bodies. Category II is an appropriate planetary protection category for missions to relatively primitive, volatile-rich, and organic-bearing small bodies that have astrobiological importance—including C-complex (C-, Cb-, Ch-, Cg-, Cgh-, and B-types), P-type, and D-type MBAs and NEOs, Trojans, comets, KBOs, and Centaurs. These objects have the potential to provide insights about prebiotic chemistry. Category II requires the provision of information that is important for future missions to the same targets, such as spacecraft impact or landing sites. The chemistry of other types of small bodies is likely not of astrobiological interest, and Category I is an appropriate category for missions to these objects, including rocky, metamorphosed, and metallic NEOs and MBAs.

Finding 4: Current scientific knowledge regarding some large asteroids (e.g., low-albedo objects ≳100 km in diameter and having an orbital semi-major axis greater than ~2.5 AU) is not sufficient to support well-informed categorization of missions to those objects, but Category II is acceptable until future reassessment. Ceres is a notable example of a large object with recently discovered importance to astrobiology and thus future missions to Ceres merit reassessing in terms of planetary protection categorization. Future missions to Ceres will likely require more rigorous planetary protection protocols than Category II.

Finding 5: The committee endorses the periodic reassessment of the planetary protection categorization scheme for all small bodies on a regular cadence. This would allow for the most recent science information to be taken into account.

Finding 6: Under current NASA and COSPAR planetary protection guidelines, Category II missions require only a minimal level of documented information, primarily target and impact/landing site.

Finding 7: Access to information prepared in response to planetary protection requirements is important for planning future missions to certain small bodies to study chemical evolution and the origin of life. The committee was unable to confirm that an archive of planetary protection information currently exists.

Finding 8: The application of planetary protection policies to private-sector space activities targeting small solar system bodies remains compromised by (1) misperceptions in the private sector about

2

planetary protection requirements; and (2) confusion about the U.S. government's ability to apply and enforce planetary protection policies concerning nongovernmental space activities.

In summary, some small solar system bodies are likely to contain prebiotic organic compounds that are relevant to the study of the origin of life in the solar system. However, based on current knowledge, it is highly improbable that small bodies harbor extinct or extant life, or that terrestrial microbes carried by a landing spacecraft can proliferate there. Therefore, the committee finds that, with the exception of large objects typified by the special case of Ceres, the forward contamination of small bodies is not expected to have a tangible effect on broader astrobiological investigations in the solar system. Nevertheless, given the importance of some classes of relatively primitive, volatile-rich, and organic-bearing small bodies to studies of prebiotic chemistry and the sparsity of current knowledge about them, the committee sees no reason to reduce the current categorizations (from Category II to Category I) for missions to such objects until such time as scientific knowledge changes. On the other hand, Category I is appropriate for missions to rocky, metamorphosed NEOs and MBAs. The committee strongly endorses periodic reassessments (on a timescale commensurate with the pace of new small-body science investigations) of the appropriate categorization of missions to all classes of objects as new scientific information becomes available. Such new or more detailed information could provide a basis for relaxing requirements for missions to some objects, or in the case of unique objects like Ceres, future mission requirements might need to be made more stringent. Finally, the community would benefit greatly from an archive of relevant mission planetary protection documentation and clearly defined lines of authority in terms of U.S. government planetary protection policy as it applies to nongovernmental space activities, and the committee strongly endorses the findings of prior reports on this topic.

1

Introduction

STUDY BACKGROUND

Planetary protection has two goals; namely, to protect the biological integrity of other solar system bodies for future science missions (preventing forward contamination)[1] and to preserve the integrity of Earth's biosphere (preventing back contamination). These goals are incorporated into National Aeronautics and Space Administration (NASA) policy,[2] international law,[3] and international scientific consensus guidance from the Committee on Space Research (COSPAR) of the International Council for Scientific Unions.[4] The NASA and COSPAR policies include specific guidelines for different types of solar system exploration missions (e.g., flybys, orbiters, landers) and for different solar system bodies (i.e., planets, asteroids, comets, etc.).

For most of the history of the space age, missions to solar system bodies beyond Earth have been conducted by a few government agencies for purely scientific purposes. That situation is changing now, especially in the sense that nongovernmental entities, including private-sector enterprises, are preparing to conduct missions to other solar system bodies for diverse reasons, such as accessing and using extraterrestrial resources. Changes such as these, as well as recent advances in the scientific understanding of solar system bodies, have led NASA and the international scientific community to conclude that reassessments of planetary protection policies are timely and important.

The Committee on Planetary Protection (CoPP) of the Space Studies Board (SSB) has responded to the need to reassess planetary protection policy implications in a 2020 report regarding lunar lander missions[5] and a 2021 report regarding Mars robotic lander missions.[6] The present CoPP report extends the committee's assessments to missions to small solar system bodies.

PLANETARY PROTECTION POLICY FOR SMALL SOLAR SYSTEM BODIES

Both NASA and COSPAR policies provide for assigning each solar system exploration mission to a specific planetary protection category. For missions that are not intended to return samples to Earth, these

---

[1] In planetary protection policy, the goal of preventing forward contamination focuses exclusively on avoiding interference with searches for evidence of life or the origin of life, and it does not encompass other esthetic, environmental, or ethical issues relating to contamination of solar system bodies.

[2] NASA, 2021, "NASA Procedural Requirements: Planetary Protection Provisions for Robotic Extraterrestrial Missions," NPR 8715.24, September 24.

[3] "Treaty on Principles Governing the Activities of States in the Exploration and Use of Outer Space, including the Moon and Other Celestial Bodies," 18 U.S.T. 2410, 610 U.N.T.S. 205, opened for signature January 27, 1967.

[4] Committee on Space Research (COSPAR), 2021, "The COSPAR Policy on Planetary Protection," approved by the COSPAR Bureau on June 3, 2021, https://cosparhq.cnes.fr/assets/uploads/2021/07/PPPolicy_2021_3-June.pdf.

[5] National Academies of Sciences, Engineering, and Medicine (NASEM), 2020, *Report Series: Committee on Planetary Protection: Planetary Protection for the Study of Lunar Volatiles*, Washington, DC: The National Academies Press, https://doi.org/10.17226/26029.

[6] NASEM, 2021, *Report Series: Committee on Planetary Protection: Evaluation of Bioburden Requirements for Mars Missions*, Washington, DC: The National Academies Press, https://doi.org/10.17226/26336.

range from Category I—for missions to objects for which there are no concerns about planetary protection—to Category IV—for missions to bodies for which there is a significant concern that contamination could be harmful to scientific investigations of evidence of life on the target body. There is a Category V for missions that will return samples from solar system bodies back to Earth.

The current policy places missions to undifferentiated, metamorphosed asteroids in Category I, for which there are no planetary protection requirements beyond identifying the target of the mission. NASA assigns missions to other asteroids and to Kuiper Belt objects to Category II, for which there are only information requirements for planetary protection.[7] COSPAR specifies that missions to carbonaceous asteroids and the dwarf planet Ceres also should fall in Category II.[8] COSPAR also recommends that forward contamination controls for small bodies other than those cited above may not be necessary and that those missions should fall into Category I or II, because there are so many members of each class of the other solar system objects. These planetary protection policies for small-body missions mean that missions are not required to comply with spacecraft cleanliness protocols or provide spacecraft organic inventories.

The committee discusses the distinctions between Categories I and II for small-body missions, and the implications of the committee's findings, in more detail in Chapter 4.

RELEVANT PRIOR REPORTS

There are several past advisory reports that are relevant to the committee's assessment of planetary protection policy for missions to small bodies, and they are summarized below.

The 1998 SSB report *Evaluating the Biological Potential in Samples Returned from Planetary Satellites and Small Solar System Bodies: Framework for Decision Making*[9] focused on recommendations for planetary protection provisions to prevent back contamination from sample-return missions and did not address forward contamination. However, the report did consider the likelihood of finding living organisms on small solar system bodies, and those conclusions are relevant to motivating forward contamination policies as well.

With regard to asteroids, the report concluded:

For samples returned from C-type asteroids, undifferentiated metamorphosed asteroids, and differentiated asteroids, the potential for a living entity in a returned sample is extremely low, but the task group could not conclude that it is zero.

For comets, the report concluded:

It is extremely unlikely that life could exist on comets, but only in a few cases can the possibility be totally ruled out, such as in the outer layers of Oort Cloud comets entering the solar system for the first time.... The task group concluded that cometary nuclei are unlikely to contain organisms capable of self-replication.

These conclusions are consistent with the present NASA and COSPAR guidelines for small-body missions.

---

[7] The characteristics of these different objects are discussed in Chapter 2.

[8] COSPAR, 2021, "The COSPAR Policy on Planetary Protection," approved by the COSPAR Bureau on June 3, 2021, https://cosparhq.cnes.fr/assets/uploads/2021/07/PPPolicy_2021_3-June.pdf.

[9] National Research Council, 1998, *Evaluating the Biological Potential in Samples Returned from Planetary Satellites and Small Solar System Bodies: Framework for Decision Making*, Washington, DC: National Academy Press, https://doi.org/10.17226/6281.

The 2019 SSB report *An Astrobiology Strategy for the Search for Life in the Universe*[10] reviewed the field of astrobiology from the perspective of relationships between searches for evidence of life in the solar system and the study of extrasolar planetary systems. The report noted the scientific synergy between understanding the role of small bodies in delivery of volatiles and organics necessary for the origin of life to the early Earth and other solar system planets and understanding similar processes for other evolving planetary systems. Thus, the report highlighted the importance of small bodies in prebiotic chemistry in the solar system.

In 2019, NASA's Planetary Protection Independent Review Board (PPIRB) report[11] included a minor recommendation that called for review of planetary protection categorization for small-body missions, as follows:

> In cases of missions to Solar System destinations where there is a large population of similar Category I and II objects (e.g., comets, asteroids, Kuiper Belt objects), NASA should allow classification of individual objects as Category I to simplify missions to them.

The PPIRB report explained that:

> In the case of small bodies where there are numerous potential targets, the contamination of any individual does not cause significant contamination to the class as a whole. If chemical evolution or origin of life experiments are planned for such objects, there are myriad to choose from that will not have been previously visited by robotic probes.

This recommendation reflects a concern that meeting planetary protection requirements for small-body missions might be unnecessarily complex or costly. The present CoPP report provides an in-depth scientific assessment of that PPIRB recommendation for small bodies.

The 2022 decadal survey for planetary science and astrobiology, *Origins, Worlds, and Life*,[12] recommended eight high-priority mission themes for consideration in the next round of competitions for NASA's medium-class New Frontiers missions. Three of those priorities addressed the science of small bodies, as follows:

- Centaur Orbiter and Lander to orbit and land on an ice-rich planetesimal in the outer solar system,
- Ceres Sample Return to collect and return samples from salt deposits on the dwarf planet Ceres, and
- Comet Surface Sample Return to map the nucleus of a Jupiter family comet and bring a sample of the surface back to Earth.

All three mission concepts would emphasize studies to understand small bodies as reservoirs of prebiotic material and potential environments that may have played a role in the origin of life in the solar system.

CHARGE TO THE COMMITTEE AND STRUCTURE OF THIS REPORT

As a standing discipline committee of the SSB to advise NASA on important matters relating to planetary protection, the committee's charge includes the preparation of short assessment reports detailing progress in areas relating to NASA's planetary protection guidelines or new scientific and technical

---

[10] NASEM, 2019, *An Astrobiology Strategy for the Search for Life in the Universe*, Washington, DC: The National Academies Press, https://doi.org/10.17226/25252.

[11] Planetary Protection Independent Review Board (PPIRB), 2019, *NASA Planetary Protection Independent Review Board (PPIRB): Report to NASA/SMD: Final Report*, Washington, DC: NASA, p. 13.

[12] NASEM, 2022, *Origins, Worlds, and Life: A Decadal Strategy for Planetary Science and Astrobiology 2023–2032*, Washington, DC: The National Academies Press, https://doi.org/10.17226/26522.

developments. Such reports may include consensus findings and discussion of the basis for the committee's conclusions, but short reports do not include specific, formal recommendations. The Statement of Task is as follows:

> The Committee on Planetary Protection (CoPP) of the Space Studies Board (SSB) shall conduct a study on planetary protection categorization of outbound-only missions to small bodies that addresses the following topics. In what follows, an "identifiable population" of solar system small bodies refers to a subset of solar system small bodies defined by ranges of measurable known parameters, such as (a) orbital elements, (b) spectroscopic classification, (c) activity, (d) composition, and/or (e) size. Objects yet to be discovered, whose properties fall into the defining ranges, are to be considered members of the corresponding identifiable population.
>
> 1. Are there identifiable populations of solar system small bodies that are sufficiently numerous, of sufficiently similar accessibility, and/or of sufficiently low relevance to the study of chemical evolution related to the search for extraterrestrial life that the contamination of *one* object in the population would reasonably be expected to have no tangible effect on the potential for scientific investigation using *other* objects in the population? If so, provide a list of these identifiable populations and their defining parameters;
> 2. For the populations identified in #1, is it appropriate to categorize all missions to objects in these as planetary protection Category I?
> 3. If, after the publication of the study, new information indicates that a previously defined identifiable population is sufficiently inhomogeneous with regard to planetary protection to warrant reassessment, what protocols should be followed in order to revise the defining parameter ranges and corresponding planetary protection categorizations?
>
> The implications of the report findings will be consistent with the budget limitations provided by NASA at the time of study initiation. The study will gather input from stakeholders, including the planetary and astrobiology science communities, government agencies dealing with spaceflight and exploration, and the aerospace industry, including emerging commercial entities.

This report responds to NASA's request. The report is based on expert briefings and discussions during the committee's five virtual open meetings from December 2021 through April 2022 and the committee's review of publicly available material.

In Chapter 2, the committee provides an overview of some key characteristics of the different types and populations of small solar system bodies, including information regarding their possible relevance to astrobiological studies.[13] Chapter 2 also presents the committee's findings about the habitability of small bodies and the risk of terrestrial microbial contamination of small bodies. Chapter 3 discusses the committee's criteria for assessing the extent to which missions to different small bodies need to implement planetary protection protocols, or not, and the chapter presents the committee's specific findings on mission categorization. Chapter 4 delves into the distinctions between Category I and II missions and the implications of those distinctions in terms of mission complexity, cost, and objectives. Chapter 5 highlights issues about implications for private-sector missions to small bodies and about a lack of clarity regarding government policy for implementing planetary protection in private-sector missions.

---

[13] There are a few types of small bodies that the committee did not consider in its analysis, including Trans-Neptunian objects at heliocentric distances beyond 50 AU and dwarf planets in the Kuiper Belt. Planetary protection assessment for missions to Pluto and other Kuiper Belt object dwarf planets will no doubt need to be considered once such a mission is prioritized; the committee anticipates that any such mission will be a scientific mission and thus scientific requirements will likely outweigh planetary protection cleanliness requirements. Finally, the committee did not consider either missions to planetary satellites or planetary defense missions designed to deflect or destroy an asteroid that might become a hazard to Earth. These were considered to be out of the scope of this particular study.

2

Small Bodies: Background and Considerations

Since the 1801 discovery of Ceres, more than 1.2 million small bodies have been discovered in the solar system.[1] These small bodies can be sorted according to their orbital characteristics into the following groups: Main Belt, near-Earth, and Trojan asteroids; comets sourced from the Kuiper Belt and the Oort Cloud; Centaurs; and Trans-Neptunian objects. These groups are further subdivided based on additional dynamical and also spectroscopic/color characteristics.

A relatively small number of small bodies have been visited by spacecraft, greatly enhancing scientific understanding of these widely varying objects, but most information is gleaned from multi-wavelength telescopic imaging and spectroscopy from Earth- and space-based facilities. Laboratory studies of meteorites and returned samples have added further knowledge. Studies over the last 200+ years have revealed sizes, densities, compositions, space weathering effects, and dynamical and cratering histories of these worlds, delivering a rich catalogue that provides evidence of the history of the solar system, including pointers to the ultimate sources of water and organics on Earth. Compositional gradients among small bodies across the solar system are linked to early solar system processes and provide clues about the building blocks of life and their distribution throughout the solar system.

DYNAMICAL GROUPINGS OF SMALL BODIES

Small solar system bodies comprise a wide variety of types of objects. Here, the committee initially discusses these objects as grouped by their dynamical configurations.

Main Belt Asteroids

Main Belt asteroids (MBAs) (see Figures 2-1 and 2-2) are those bodies in heliocentric orbits with semi-major axes between the orbits of Mars and Jupiter. Ceres is particularly notable as the largest object in the Main Belt (940 km in diameter) and was classed as a dwarf planet in 2006. (See Box 2-1 on Ceres for more information.) Rare, giant collisions among asteroids have resulted in dynamical asteroid families, wherein gravitational re-accumulation of material after a large collision leads to the formation of an entire family of large and small objects with dynamical properties similar to those of the original body.[2] Many asteroid families may have dispersed since formation and are thus difficult to identify in the current era.[3]

---

[1] IAU Minor Planet Center, 2022, "Running Tallies," http://www.minorplanetcenter.net.

[2] P. Michel, 2001, "Collisions and Gravitational Reaccumulation: Forming Asteroid Families and Satellites," *Science* 294(5547):1696–1700, https://doi.org/10.1126/science.1065189.

[3] D. Nesvorný, M. Broz, and V. Carruba, 2015, "Identification and Dynamical Properties of Asteroid Families," Pp. 297–321 in *Asteroids IV*, W.F. Bottke, F.E. DeMeo, and P. Michel, eds., Tucson: University of Arizona Press, https://doi.org/10.2458/azu_uapress_9780816532131-ch016.

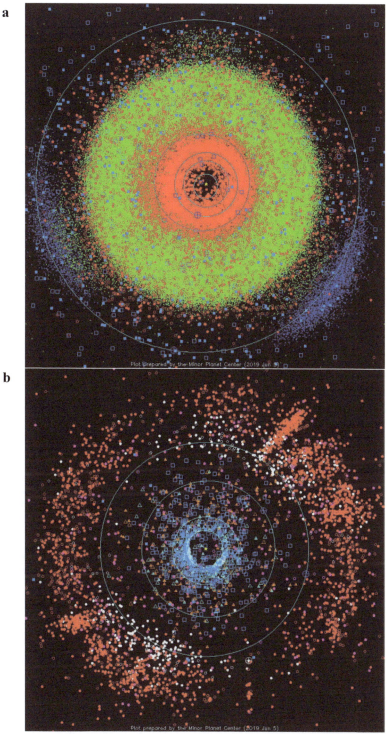

FIGURE 2-1 (a) View of small bodies in the inner solar system (June 5, 2019); vernal equinox is to the right. Outer blue circle represents Jupiter's orbit. Green symbols: numbered minor planets (asteroids), red: NEOs, deep blue: Jupiter Trojans, light blue: comets. (b) View of small bodies in the outer solar system (June 5, 2019); vernal equinox is to the right. Jupiter's orbit is represented by the innermost blue circle. Orange: Centaurs; red: "classical" Kuiper Belt objects; light-blue: comets.
SOURCE: Minor Planet Center, The Center for Astrophysics | Harvard & Smithsonian. Licensed under CC0 from https://www.minorplanetcenter.net/iau/lists/InnerPlot.html.

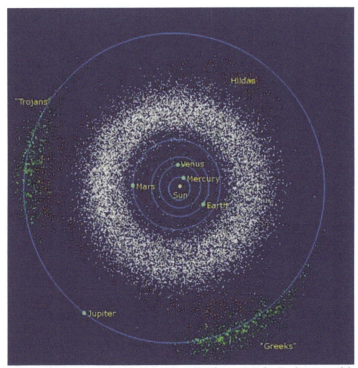

FIGURE 2-2 View of Main Belt asteroids and Jupiter Trojans. Main Belt asteroids shown in white (including Hildas, orange) and Trojans, green. The Hildas (~3.7–4.2 AU) are in 3:2 mean motion resonance with Jupiter.
SOURCE: Wikipedia File:InnerSolarSystem-en.png, public domain, Wikipedia user Mdf.

Collisions among asteroids can also result in the formation of natural satellites,[4,5] observations of which can help constrain the density of the parent asteroid.[6] Models of the collisional history within the asteroid belt, using the current size frequency distribution as a constraint, suggest that asteroids of diameter $\gtrsim$100 km are primordial, with their physical properties likely determined during the accretion epoch[7] (Vesta is an exception here given its differentiated nature).[8] Most smaller asteroids ($\lesssim$40 km diameter) are byproducts of fragmentation events. The Hildas (shown in Figure 2-2) are asteroids at the outer part of the Main Belt (~3.7–4.2 AU) in 3:2 mean motion resonance with Jupiter; these objects are predicted to have originated in the same region as the Trojans and Kuiper Belt objects (KBOs) (beyond the primordial orbits of the ice giants).

[4] D.D. Durda, W.F. Bottke, B.L. Enke, et al., 2004," The Formation of Asteroid Satellites in Large Impacts: Results from Numerical Simulations," *Icarus* 170(1):243–257, https://doi.org/10.1016/j.icarus.2004.04.003.

[5] The first asteroidal satellite discovered was Dactyl, the moon of (243) Ida, the target of a flyby of the Galileo spacecraft in 1993. M.J.S. Belton, C.R. Chapman, P.C. Thomas, et al., 1995, "Bulk Density of Asteroid 243 Ida from the Orbit of Its Satellite Dactyl," *Nature* 374(6525):785–788, https://doi.org/10.1038/374785a0.

[6] W.J. Merline, L.M. Close, C. Dumas, et al., 1999, "Discovery of a moon orbiting the asteroid 45 Eugenia," *Nature* 401(6753):565–568, https://doi.org/10.1038/44089.

[7] W.F. Bottke, D.D. Durda, D. Nesvorný, et al., 2005, "The Fossilized Size Distribution of the Main Asteroid Belt," *Icarus* 175(1):111–140, https://doi.org/10.1016/j.icarus.2004.10.026.

[8] A. Ruzicka, G.A. Snyder, and L.A. Taylor, 1997, "Vesta as the Howardite, Eucrite and Diogenite Parent Body: Implications for the Size of a Core and for Large-Scale Differentiation," *Meteoritics and Planetary Science* 32:825–840, https://doi.org/10.1111/j.1945-5100.1997.tb01573.x.

BOX 2-1 Ceres: A Case Study

Ceres is unique among inner solar system bodies. Little was known about Ceres prior to the arrival of NASA's Dawn mission, as its surface appeared in ground-based observations to be relatively homogenous, and because of the lack of meteorites identified as originating from Ceres. Dawn greatly expanded knowledge of Ceres, owing to the multi-instrument observations from a range of orbital altitudes. Ceres has thus advanced from a point of light to an astrobiologically interesting target as a result of the Dawn mission.

Ceres is now known to be the most water-rich body in the inner solar system after Earth. Ceres's visible-near infrared spectrum is similar to some C-complex asteroids, but its bulk density of 2,162 kg m^{-3},[a] intermediate between water ice and silicates, is more similar to icy moons of the outer solar system, and even early measurements showed the presence of hydrated minerals on the surface.[1] Early thermal evolution models of Ceres indicate that it may have differentiated into a silicate core and a water-rich outer layer, which could suggest that Ceres harbored a global subsurface ocean for several hundred million years after its formation.[b] Measurements of hydrogen by Dawn indicate the presence of a global, subsurface water-ice table at depths less than a few decimeters at latitudes greater than 45°.[c] Water ice is also associated with some impact craters on Ceres. The Herschel Space Observatory detected water vapor around Ceres,[d] though subsequent measurements, including by Dawn instruments, have not confirmed this and may point to a transient exosphere.

Dawn revealed Ceres to exhibit extensive chemical and geological brine-driven activity, likely occurring until relatively recently and possibly even currently. Evidence of cryovolcanism and more than 100 bright spots (faculae) are seen across the Cerean surface; the faculae, 90 percent of which are related to relatively young impact craters, are found to be enriched in sodium carbonates and ammonium salts. High-resolution observations of the largest craters show evidence of recent cryovolcanic activity with salts potentially still making their way to the surface. Dawn also revealed localized spots rich in organic material.[e]

Ceres's geologic activity was unexpected because, unlike the icy moons of the outer solar system, Ceres is not heated by tidal interactions with a large planet and was expected to be cold and inactive. However, Ceres shares many similarities with the ocean worlds of the outer solar system planets, and leading to Ceres now being considered a candidate ocean world,[f] likely a key endmember that can help to understand the evolution of ocean worlds. Ceres is the most easily accessible ocean world candidate and potentially important to the study of astrobiology. Ceres's similarity with outer solar system ocean worlds, along with Dawn's detection of ammoniated phyllosilicates, spurs the question: did Ceres form in the outer solar system and migrate inward? Future missions to Ceres will probe this and other questions that have stemmed from Dawn observations.

FIGURE 2-1-1 *Left:* Ceres as seen by Hubble Space Telescope and Dawn. *Right:* Hubble Space Telescope images of Ceres (right).
NOTE: Enhanced color image of Ceres from Dawn; the bright spot near center is the salt-rich facula associated with Occator crater.
SOURCES: *Left:* NASA, ESA, J. Parker (Southwest Research Institute), P. Thomas (Cornell University), L. McFadden (University of Maryland, College Park), and M. Nutchler and Z. Levay (STScI). *Right:* NASA/JPL-CalTech/UCLA/MPS/DLR/IDA.

a J.C. Castillo-Rogez, et al., 2020, "Ceres: Astrobiological Target and Possible Ocean World," *Astrobiology* 20, https://doi.org/10.1089/ast.2018.1999; L.A. Lebofsky, 1978, "Asteroid 1 Ceres: Evidence for Water of Hydration," *Monthly Notices of the Royal Astronomical Society* 182:17P–21P, https://doi.org/10.1093/mnras/182.1.17P.

b T.B. McCord and C. Sotin, 2005, "Ceres: Evolution and Current State," *Journal of Geophysical Research (Planets)* 110(E5), https://doi.org/10.1029/2004JE002244; J.C. Castillo-Rogez and T.B. McCord, 2010, "Ceres' Evolution and Present State Constrained by Shape Data," *Icarus* 205(2):443–459, https://doi.org/10.1016/j.icarus.2009.04.008.

c T.H. Prettyman, et al., 2017, "Extensive Water Ice Within Ceres' Aqueously Altered Regolith: Evidence from Nuclear Spectroscopy," *Science* 355:55–59, https://doi.org/10.1126/science.aah6765.

d M. Küppers, et al., 2014, "Localized Sources of Water Vapour on the Dwarf Planet (1) Ceres," *Nature* 505:525–527, https://doi.org/10.1038/nature12918.

e T.B. McCord, et al., 2021, "Ceres, a Wet Planet: The View After Dawn," *Geochemistry*, https://doi.org/10.1016/j.chemer.2021.125745; J.C. Castillo-Rogez, M. Neveu, H.Y. McSween, R. Fu, M.J. Toplis, and T.H. Prettyman, 2018, "Insights into Ceres' Evolution from Surface Composition," *Meteorics and Planetary Science* 53:1820–1843, https://doi.org/10.1111/map.13181; H.Y. McSween, J.P. Emery, A.S. Rivkin, M.J. Toplis, J.C. Castillo-Rogez, T.H. Prettyman, M.C. De Sanctis, C.M. Pieters, C.A. Raymond, and C.T. Russell, 2018, "Carbonaceous Chondrites as Analogs for the Composition Alteration of Ceres," *Meteorics and Planetary Science* 53:1793–1804, https://doi.org/10.1111/maps.13124.

f A.R. Hendrix, et al., 2019, "The NASA Roadmap to Ocean Worlds," *Astrobiology* 19:1–27, https://doi.org/10.1089/ast.2018.1955; J.C. Castillo-Rogez, et al., 2020, "Ceres: Astrobiological Target and Possible Ocean World," *Astrobiology* 20(2), https://doi.org/10.1089/ast.2018.1999.

Near-Earth Objects

Near-Earth objects (NEOs, or near-Earth asteroids, NEAs) are those objects with perihelia within 1.3 AU.[9] These objects originate primarily in the main asteroid belt and are ejected into near-Earth space via dynamical "escape hatches," whereby asteroid fragments are constantly created by both collisions and mass shedding events. A fraction of this population of fragments, namely those of diameters $\lesssim$30 km, can escape the Main Belt via the gravitational resonances, thereby creating a quasi-steady-state population of near-Earth asteroids,[10] with a dynamical duration in the inner solar system of ~10 Myr. NEOs, as fragments of MBAs, thus represent a sample of some compositional types of MBAs that are more dynamically accessible from Earth than many MBAs, making them attractive targets for sample-return missions such as Hayabusa, Hayabusa2, and OSIRIS-Rex [Origins, Spectral Interpretation, Resource Identification, Security, Regolith Explorer]. With orbits in the inner solar system, NEOs experience greater amounts of processing (e.g., extreme heating, solar wind exposure) and consequently, their surfaces are less pristine than their parent bodies in the Main Belt. As evidenced by Ryugu and Bennu (targets of Hayabusa2 and OSIRIS-REx, respectively), however, carbonaceous NEOs do retain some level of volatiles.

Kuiper Belt Objects and Centaurs

KBOs (a subclass of Trans-Neptunian objects, TNOs) are small bodies inhabiting the outer reaches of the solar system, extending from roughly 30 to 50 AU; the region is named for Gerard Kuiper, who speculated some seven decades ago about objects beyond Pluto. The so-called cold classical KBOs have orbits with relatively low inclinations and eccentricities, while the orbits of hot classical KBOs are more highly inclined and less circular, having been more influenced and perturbed by Neptune's gravity. Dwarf planet Pluto is known as one of the largest KBOs, and the contact binary KBO Arrokoth[11] was the target

---

[9] Center for Near Earth Object Studies, "NEO Basics," https://cneos.jpl.nasa.gov/about/neo_groups.html.

[10] W.F. Bottke, M. Brož, D.P. O'Brien, A.C. Bagatin, A. Morbidelli, and S. Marchi, 2015, "The Collisional Evolution of the Main Asteroid Belt," Pp. 701–724 in *Asteroids IV*, P. Michel, F.E. DeMeo, and W.F. Bottke, eds., Tucson: University of Arizona Press.

[11] W.M. Grundy, M.K. Bird, D.T. Britt, et al., 2020, "Color, Composition, and Thermal Environment of Kuiper Belt Object (486958) Arrokoth," *Science* 367(6481), https://doi.org/10.1126/science.aay3705.

of the New Horizons flyby in 2019. The Nice model[12] describes perturbations of the primordial belt as including resonant capture and scattering of KBOs by an outward migrating Neptune; even after its migration ended, Neptune has continued to erode the Kuiper Belt by gravitational scattering, by sending objects outward (to the "scattered disk") or inward to become Centaurs and precursors to the Jupiter family comets discussed below. The orbits of Centaurs occupy the space between the orbits of Jupiter and Neptune, interacting strongly with the gravity of these giant planets and as a result are either ejected from the solar system or pushed into the inner solar system where they become comets.

Jupiter Trojans

Jupiter Trojans are those bodies that orbit the Sun near the stable Jupiter Lagrangian points L4 and L5, leading and trailing Jupiter by 60°. The number of asteroids in the leading group is larger than that of the trailing group, by a factor of ~1.4±0.2 for Trojans larger than 10 km.[13] Orbital inclinations of these objects are up to 40°.[14] In the Nice model, it is proposed that resonant interactions between Jupiter and Saturn temporarily destabilized the orbits of Uranus and Neptune, which moved into the primordial Kuiper Belt, scattering material widely across the solar system. In this model, Jupiter's primordial Trojan population was lost and the Lagrange regions were repopulated with this scattered Kuiper Belt material.[15] The Jupiter Trojans thus may represent KBOs currently orbiting the Sun at 5.2 AU. The Lucy mission will make the first up-close observations of Trojans during flybys of five of these objects (including one binary pair) in 2027–2033.

Comets

Comets are volatile-rich bodies that can periodically enter the inner solar system following long-term storage in the Kuiper Belt and the Oort Cloud reservoirs. Whereas asteroids largely have low-eccentricity orbits (mostly < ~0.3–0.4), most comets have higher-eccentricity orbits. The short-period comets (Jupiter Family comets, JFCs), with periods of 5–10 years, tend to have lower inclinations and are thought to originate in the Kuiper Belt (see Figure 2-3). Jan Oort first suggested that the long-period comets (P > 200 years) that enter the inner solar system come from a cloud of icy bodies as far as 2,000 to 100,000 AU from the Sun.[16] The Oort cloud is estimated to hold billions or even trillions of bodies; when these objects in the cloud interact with passing stars, molecular clouds, and gravity from the galaxy, they can spiral inward toward the Sun as long-period comets.[17,18]

---

[12] A. Morbidelli, H.F. Levison, K. Tsiganis, and R. Gomes, 2005, "Chaotic Capture of Jupiter's Trojan Asteroids in the Early Solar System," *Nature* 435:462–465, https://doi.org/10.1038/nature03540.

[13] T. Grav, A.K. Mainzer, J. Bauer, et al., 2011, "WISE/NEOWISE Observations of the Jovian Trojans: Preliminary Results," *The Astrophysical Journal* 742(40), https://doi.org/10.1088/0004-637X/742/1/40.

[14] S. Pirani, A. Johansen, and A.J. Mustill, 2019, "On the Inclinations of the Jupiter Trojans," *Astronomy and Astrophysics* 631:A89, https://doi.org/10.1051/0004-6361/201936600.

[15] E. Dotto, J.P. Emery, M.A. Barucci, A. Morbidelli, and D.P. Cruikshank, 2008, "De Troianis: The Trojans in the Planetary System," Pp. 383–395 in *The Solar System Beyond Neptune*, M.A. Barucci, H. Boehnhardt, D.P. Cruikshank, and A. Morbidelli, eds., Tucson: University of Arizona Press.

[16] J.H. Oort, 1950, "The Structure of the Cloud of Comets Surrounding the Solar System and a Hypothesis Concerning Its Origin," *Bulletin of the Astronomical Institutes of the Netherlands* 11(408):91–110, https://hdl.handle.net/1887/6036.

[17] R. Brasser and A. Morbidelli, 2013, "Oort Cloud and Scattered Disc Formation During a Late Dynamical Instability in the Solar System," *Icarus* 225:40–49, https://doi.org/10.1016/j.icarus.2013.03.012.

[18] R. Brasser and M.E. Schwamb, 2015, "Re-Assessing the Formation of the Inner Oort Cloud in an Embedded Star Cluster—II. Probing the Inner Edge," *Monthly Notices of the Royal Astronomical Society* 446:3788–3796, https://doi.org/10.1093/mnras/stu2374.

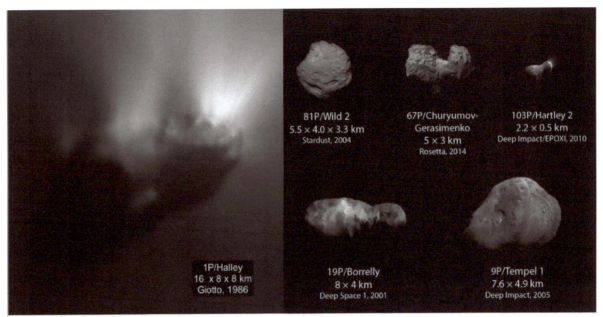

FIGURE 2-3 Diversity among Jupiter family comets.
NOTE: 26P Grigg-Skjellerup is not included because images were not captured during that flyby.
SOURCES: *Halley:* Russian Academy of Sciences/T. Stryk. *Borrelly:* NASA/JPL/Ted Stryk. *Tempel 1 and Hartley 2:* NASA/JPL/UMD. *Churyumov-Gerasimenko:* ESA/Rosetta/NavCam/E. Lakdawalla. *Wild 2:* NASA/JPL. Montage by E. Lakdawalla. CC BY-NC-ND 3.0.

COMPOSITIONAL, TAXONOMIC, AND SPECTRAL DISTINCTIONS

Much of what is known about small bodies' compositions comes from Earth-based spectroscopic observations. MBAs and NEOs are grouped taxonomically by their spectral properties at visible and near-infrared (NIR) wavelengths (~0.4–2.4 microns) (Figure 2-4). The first asteroid taxonomy was assembled by Chapman, Morrison, and Zellner[19] in 1975. Since then, various additional taxonomic systems have been published based on broadband colors covering the wavelength range 0.337–1.055 microns coupled with visible albedo to delineate asteroids,[20] as well as using narrowband spectroscopy in the 0.44–0.92 micron spectral range[21] and through longer NIR wavelengths.[22] In all of these systems, S-types are rocky (melted/metamorphosed); C-complex (including B, C, Cg, Cb, Ch, and Cgh) are low-albedo and presumably more carbonaceous and volatile-rich (aqueously altered); and P- and D-types are low-albedo and spectrally redder, likely consistent with an organic-rich composition. M-type asteroids are thought to be metal-rich. Additional taxonomic types are discussed in the literature but are not included here. These various taxonomic types are distributed throughout the Main Belt, though important trends are observed (Figure 2-5). In particular, the majority of the mass of the inner Main Belt is dominated by S-type, rocky material, while the mass of the outer belt is dominated by more volatile-rich, carbonaceous asteroids.

---

[19] C.R. Chapman, D. Morrison, and B. Zellner, 1975, "Surface Properties of Asteroids: A Synthesis of Polarimetry, Radiometry, and Spectrophotometry," *Icarus* 25:104–130, https://doi.org/10.1016/0019-1035(75)90191-8.

[20] D.J. Tholen and M.A. Barucci, 1989, "Asteroid Taxonomy," Pp. 298–315 in *Asteroids II*, R.P. Binzel, T. Gehrels, M.S. Matthews, eds., Tucson: University of Arizona Press.

[21] S.J. Bus and R.P. Binzel, 2002, "Phase II of the Small Main-Belt Asteroid Spectroscopic Survey: A Feature-Based Taxonomy," *Icarus* 158:146–177, https://doi.org/10.1006/icar.2002.6856.

[22] F.E. DeMeo, R.P. Binzel, S.M. Slivan, and S.J. Bus, 2009, "An Extension of the Bus Asteroid Taxonomy into the Near-Infrared," *Icarus* 202:160–180, https://doi.org/10.1016/j.icarus.2009.02.005.

Based on Earth-based observations, the Jupiter Trojans are comprised of largely P- and D-type asteroids. A long-standing paradigm is that the low-albedo and red spectral slopes are due to the presence of complex organic molecules.[23] Based on reported signatures of fine-grained silicates[24] on some Trojans, a bulk-density measurement,[25] and their locations at 5.2 AU, Trojans have generally been inferred to contain a large fraction of H_2O ice, though covered by a refractory mantle, and a higher abundance of complex organic molecules than most MBAs. The Hildas at the outermost part of the Main Belt, exhibit the same bimodal color distribution as the Trojans.[26] KBOs are generally classed into "red" and "less-red" groupings based on their spectral slopes at visible wavelengths.

As of October 2021, a total of 27,196 NEOs have been discovered, and more than 1.1 million Main Belt asteroids (>1 km in diameter) are known; 4,429 comets have been discovered (as of December 2021; about 3,000 of these belong to the Kreutz family of Sun-grazing comets), and there are ~9,800 known Jupiter Trojans. Given these numbers, even a body with a rare taxonomic type is likely represented by innumerable smaller, similar bodies. However, all bodies of the same "type" cannot be assumed to be the same; as discussed later, spacecraft visits have played important roles in studying diversity among apparently taxonomically similar asteroids.

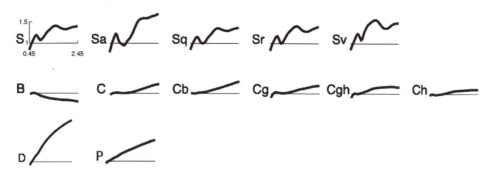

FIGURE 2-4 Representation of the visible-near infrared spectral shapes of the major taxonomic types of asteroids discussed in this report. The x axis represents wavelengths of 0.45–2.45 microns while the y axis represents normalized reflectance of values 1–1.5. *Top row:* S-types and subclasses of S-types. *Middle row:* C-complex. *Bottom row:* D- and P-types.
SOURCE: F.E. DeMeo, R.P. Binzel, S.M. Slivan, and S.J. Bus, 2009, "An Extension of the Bus Asteroid Taxonomy into the Near-Infrared," *Icarus* 202(1):160–180, https://doi.org/10.1016/j.icarus.2009.02.005. Reprinted, Copyright 2009, with permission from Elsevier.

[23] J. Gradie and J. Veverka, 1980, "The Composition of the Trojan Asteroids," *Nature* 283:840–842, https://doi.org/10.1038/283840a0.

[24] J.P. Emery, D.P. Cruikshank, and J. Van Cleve, 2006, "Thermal Emission Spectroscopy (5.2 38 μm) of Three Trojan Asteroids with the Spitzer Space Telescope: Detection of Fine-Grained Silicates," *Icarus* 182:496–512, https://doi.org/10.1016/j.icarus.2006.01.011.

[25] F. Marchis, D. Hestroffer, P. Descamps, et al., 2006, "A Low Density of 0.8gcm3 for the Trojan Binary Asteroid 617 Patroclus," *Nature* 439:565–567, https://doi.org/10.1038/nature04350.

[26] I. Wong and M.E. Brown, 2016, "A Hypothesis for the Color Bimodality of Jupiter Trojans," *The Astronomical Journal* 152(90), https://doi.org/10.3847/0004-6256/152/4/90.

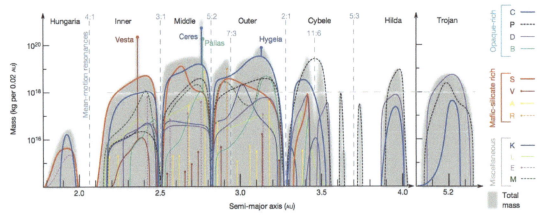

FIGURE 2-5 Spatial distribution of asteroid taxonomic types. (This graphic shows more taxonomic types of asteroids than are discussed in this report.) The distribution shows significant overlap, but melted and metamorphosed asteroids tend to be concentrated in the inner belt, aqueously altered bodies in the middle, and bodies in which ice never melted in the outer belt.
SOURCE: F.E. DeMeo and B. Carry, 2014, "Solar System Evolution from Compositional Mapping of the Asteroid Belt," *Nature* 505:629–634, https://doi.org/10.1038/nature12908. Reprinted by permission from Springer Nature, Copyright 2014.

The number of discovered NEOs is expected to increase by orders of magnitude when the Vera C. Rubin Observatory begins operations; identifications of more than 100,000 NEOs are anticipated by the Legacy Survey of Space and Time. In addition, sample returns from NEOs, and improved orbital characterizations will help fill the knowledge gaps for small-body objects of sizes smaller than 10 m. Toward this outcome, spectroscopic analyses are also being conducted to compare the asteroid spectra to the spectra of meteorites held in collections, which may provide the link to their parent families.[27] Over recent decades, about 1,000 NEOs have been observed by radar, and this has provided information on both their shape and size. Furthermore, space-based infrared measurements by the Spitzer and NEOWISE observatories and the ground-based IRTF/MIRSI [Infrared Telescope Facility/Mid-InfraRed Spectrometer and Imager] observation platforms have provided additional data on the morphology of such objects. The space science community has already acquired very detailed information on the composition of three NEOs from sample-return missions to Itokawa, Bennu, and Ryugu.

Comets, with sources in the most distant reaches of the solar system, represent the oldest relics of solar system formation and thus can serve as records of the formation period. Their compositions of carbons and other volatile species are the building blocks of planets (dust, ices, and organics).

Comets have long been studied by dedicated ground-based and space-based remote sensing campaigns. Spectroscopic and photometric studies of cometary comae provided a wealth of data on an increasing sample of comets.[28,29] These observations have led to many discoveries and have allowed for a detailed comparison between comets, other solar system objects, and have provided a link between objects in our solar system and the interstellar medium (see Box 2-2).

---

[27] R.P. Binzel, F.E. DeMeo, E.V. Turtelboom, et al., 2019, "Compositional Distributions and Evolutionary Processes for the Near-Earth Object Population: Results from the MIT-Hawaii Near-Earth Object Spectroscopic Survey (MITHNEOS)," *Icarus* 324:41–76, https://doi.org/10.1016/j.icarus.2018.12.035.

[28] M.J. Mumma, M.A. Disanti, K. Magee-Sauer, et al., 2005, "Parent Volatiles in Comet 9P/Tempel 1: Before and After Impact," *Science* 310:270–274, https://doi.org/10.1126/science.1119337.

[29] D.C. Lis, D. Bockelée-Morvan, R. Güsten, et al., 2019, "Terrestrial Deuterium-to-Hydrogen Ratio in Water in Hyperactive Comets," *Astronomy and Astrophysics* 625(L5), https://doi.org/10.1051/0004-6361/201935554.

BOX 2-2 Composition of Comets

In the context of this report, comets are important as a potential reservoir of organic molecules for the early Earth, and can also be used to study connections between our solar system and the interstellar medium (ISM). A review paper[a] discusses the origin and evolution of the material in 67P/Churyumov-Gerasimenko (C-G), which puts what we have learned about the composition of comets into the wider context of solar system evolution. In addition to the Rosetta mission, significant understanding of comets, particularly their organic matter, has been gained from samples returned by the Stardust mission and interplanetary dust particles (IDPs) (A7 REFs). The European Space Agency's Rosetta mission to comet 67P/C-G provided numerous new insights to test contemporary theories and to verify current understanding of the major processes that are involved in cometary physics:

- After the first measurement of the HDO/H_2O ratio in a Jupiter-family comet (JFC) (i.e., Hartley 2), it was hypothesized that there is an intrinsic difference between Oort cloud and Kuiper belt/scattered disk objects. This rekindled the scientific discussion on Earth's water being of cometary origin. However, recent measurements of elevated deuterium-to-hydrogen (D/H) ratios in the water of additional JFCs, including 67P/C-G, contradicted this view.[b] Thus, it seems unlikely that comets from the Kuiper belt and beyond were the major source for the water on Earth. Instead, the results suggest that the D/H ratio reflects the formation location of comets rather than where they are found today: comets formed over a wide range of heliocentric distances and were relocated afterwards.
- The refractories in other comets, including the returned samples from comet Wild 2, contain thermally processed material from the inner solar system with only small amounts of presolar material. Thus, radial mixing occurs, but the protoplanetary disk remained only partially homogenized. Therefore, the isotopic and possibly also elemental and molecular abundances in at least the icy phase at the location of the comet's formation differed from bulk solar system.
- The relative abundances of the volatiles in comets exhibit similarities to the ISM and comet 67P/C-G is no exception.[c] Furthermore, the large amounts of unsaturated hydrocarbons and other organics[d] are consistent with the ISM expectations, solidifying an ISM heritage of the volatile material in 67P/C-G.

Numerous molecular species are associated with life as we know it but have an abiotic origin in 67P/C-G and other comets. In addition to O_2, this includes organohalogens, glycine, phosphorous, and various organosulfurs. If comets did not deliver water to Earth, they could nevertheless have delivered sizeable amounts of other materials to Earth. For instance, copious amounts of organic species could also have found their way to the inner solar system and Earth.[e]

------

[a] M. Rubin, et al., 2020, "On the Origin and Evolution of the Material in 67P/Churyumov-Gerasimenko," *Space Science Reviews* 216(102), https://doi.org/10.1007/s11214-020-00718-2.

[b] L. Paganini, et al., 2017, "Ground-Based Detection of Deuterated Water in Comet C/2014 Q2 (Lovejoy) at IR Wavelengths," *Astrophysical Journal* 836:L25, https://doi.org/10.3847/2041-8213/aa5cb3.

[c] M.N. Drozdovskaya, et al., 2019, "Ingredients for Solar-Like Systems: Protostar IRAS 16293-2422 B Versus Comet 67P/Churyumov-Gerasimenko," *Monthly Notices of the Royal Astronomical Society* 490:50–79, https://doi.org/10.1093/mnras/stz2430.

[d] M. Schuhmann, et al., 2019, "Aliphatic and Aromatic Hydrocarbons in Comet 67P/Churyumov-Gerasimenko Seen by ROSINA," *Astronomy and Astrophysics* 630:A31, https://doi.org/10.1051/0004-6361/201834666.

[e] M. Rubin, et al., 2019, "Volatile Species in Comet 67P/Churyumov-Gerasimenko: Investigating the Link from the ISM to the Terrestrial Planets," *ACS Earth and Space Chemistry* 3:1792–1811.

Early remote sensing observations of the apparently anomalous acceleration of comets Encke, d'Arrest, and Wolf 1 led to the conclusion[30] that comets are "dirty snowballs" emitting gas and dust which can either decelerate or accelerate the comet along its orbit around the Sun, known as nongravitational forces. However, only in 1986 did the European Space Agency Giotto mission and the Soviet Vega 2 mission flying by comet 1P/Halley[31] confirm that comets possess a solid nucleus composed of volatile and refractory materials. From these investigations, it was established[32] that comets have preserved the accreted and condensed materials moreso than other objects in the solar system.

In general, however, though much work has been accomplished regarding small-body taxonomies and spectral and compositional studies, neither taxonomic classes nor higher resolution spectra are highly diagnostic of unambiguous composition and they generally do not reveal composition of minor constituents. Furthermore, interpretations of spectra are rendered difficult by effects such as space weathering, temperatures, and particle sizes.

LINKS TO METEORITES

In addition to spectroscopic surveys, comparisons between spectra of meteorites and asteroids can help illuminate the likely composition of parent body asteroids, particularly because space weathering processes (e.g., solar wind bombardment) can alter the spectral properties of the surfaces of asteroids, confounding the identification of the surface components.[33] Ordinary chondrite meteorites are good representatives of rocky types of asteroids larger than 10 m, and there are remarkable linkages between howardite-eucrite-diogenite meteorites and asteroid 4 Vesta.[34] Importantly, less than 5 percent of meteorite falls are carbonaceous chondrite meteorites, while their presumed parent bodies, C-complex asteroids, are plentiful in the Main Belt. Meteors from these more primitive types of asteroids may be more fragile[35] and may not survive atmospheric entry. Nevertheless, organics are present in some meteorites; they represent precursors to life, but are not indicators of life. Rather, the organic matter in comets and asteroids is derived from interstellar space,[36] as revealed by extreme isotopic fractionations, but has been further processed in the solar nebula and/or within small bodies after accretion.

Significant progress has been made connecting asteroids and meteorites. For instance, despite spectral differences, it is now understood that ordinary chondrites have S-type parent bodies whose surfaces are more space-weathered than the interiors of the meteorites. However, gaps exist in attempts to connect

[30] F.L. Whipple, 1951, "A Comet Model. II. Physical Relations for Comets and Meteors," *The Astrophysical Journal* 113(464), https://doi.org/10.1086/145416.

[31] R. Reinhard, 1986, "The Giotto Encounter with Comet Halley," *Nature* 321:313–318, https://doi.org/10.1038/321313a0.

[32] J. Geiss, 1987, "Composition Measurements and the History of Cometary Matter," *Astronomy and Astrophysics* 187:859–866.

[33] See, e.g., C.M. Pieters, L.A. Tayler, S.K. Noble, et al., 2000, "Space Weathering on Airless Bodies: Resolving a Mystery with Lunar Samples," *Meteoritics and Planetary Science* 35:1101–1107, https://doi.org/10.1111/j.1945-5100.2000.tb01496.x.

[34] H.Y. McSween, Jr., R.P. Binzel, M.C. De Sanctis, et al., 2013, "Dawn; the Vesta-HED Connection; and the Geologic Context for Eucrites, Diogenites, and Howardites," *Meteorics and Planetary Science* 48:2090–2104, https://doi.org/10.1111/maps.12108.

[35] A.L. Graps, P. Blondel, G. Bonin, et al., 2016, "ASIME 2016 White Paper: In-Space Utilisation of Asteroids: 'Answers to Questions from the Asteroid Miners,' " *arXiv* 1612.00709v2, https://doi.org/10.48550/arXiv.1612.00709.

[36] H. Busemann, A.F. Young, C.M.O'D. Alexander, P. Hoppe, S. Mukhopadhyay, and L.R. Nittler, 2006, "Interstellar Chemistry Recorded in Organic Matter from Primitive Meteorites," *Science* 312:727–730, https://doi.org/10.1126/science.1123878.

asteroids and meteorites.[37] For instance,[38] the asteroid 2008 TC3 (known as Almahata Sitta) lost more than 99 percent of its mass in the atmosphere. Its spectrum was flat, and yet a remarkable mixture of composition was found in the meteorite samples. Indeed, presumed primitive carbonaceous asteroids have relatively flat and featureless spectra that are difficult to link to specific chondrite groups.[39] Ground-based data do not provide information on possible asteroid surface and interior heterogeneities. Furthermore, dynamically distant objects (Hildas, Trojans, and Kuiper Belt) are underrepresented or absent in the meteorite collection (perhaps because these parent bodies are structurally weak as mentioned above).[40]

ICE IN THE ASTEROID BELT

A subclass of MBAs is the so-called "active asteroids."[41] Active asteroids are bodies that exhibit comet-like mass loss, ejecting volatiles and dust and producing transient, comet-like comae and tails, but have asteroid-like orbits; a few are found in near-Earth space. More than 30 active asteroids have been discovered since 1996. The hypothesized causes of activity in these bodies include impact ejection and disruption, rotational instabilities, and dehydration stresses and thermal fracture, in addition to the sublimation of asteroidal ice. Whereas "disrupted asteroids" are those active asteroids whose activity is driven by processes such as impacts and rotational disruption, the activity of another subset of active asteroids, "Main Belt comets (MBCs)," is driven by sublimation of volatiles; these MBCs thus provide new clues regarding the abundance of asteroid ice, and the origin of terrestrial planet volatiles. The activity of MBCs, driven by sublimation of volatiles, is evidence for water in the Main Belt; there are likely many more water-rich asteroids (probably mostly in the outer Main Belt) that have not yet shown activity. (Note that none of the water present in/on small bodies is liquid.) Further evidence of water comes from spectroscopic clues, on the low-albedo classes of asteroids. The 0.7 micron feature detected in some asteroids (particularly the Ch- and Cgh-types) is an oxidized iron feature indicative of Fe-bearing phyllosilicates that is the result of aqueous alteration.[42] The 3 micron feature[43] is prevalent at low-albedo asteroids and is indicative of OH or H_2O; the maximum absorption of H_2O is at 3.1 microns. Organics, with diagnostic absorptions in the 3.4–3.5 micron range, have also been found on some low-albedo asteroids.[44,45] In the Main Belt, depending on the thermal properties (e.g., composition, porosity, grain sizes, etc.) of the dust, ice can remain for a long period of time; modeling has demonstrated that

[37] C.R. Chapman, D. Morrison, and B. Zellner, 1975, "Surface Properties of Asteroids: A Synthesis of Polarimetry, Radiometry, and Spectrophotometry," *Icarus* 25:104–130, https://doi.org/10.1016/0019-1035(75)90191–90198.

[38] A.L. Graps, P. Blondel, G. Bonin, et al., 2016, "ASIME 2016 White Paper: In-Space Utilisation of Asteroids: 'Answers to Questions from the Asteroid Miners,' " *arXiv* 1612.00709v2, https://doi.org/10.48550/arXiv.1612.00709.

[39] D.S. Lauretta, O. Barnouin-Jha, M.A. Barucci, et al., 2009, "Astrobiology Research Priorities for Primitive Asteroids," Lunar and Planetary Laboratory, https://www.lpi.usra.edu/decadal/sbag/topical_wp/lauretta_etal.pdf.

[40] P. Vernazza, M. Marsset, P. Beck, et al., 2015, "Interplanetary Dust Particles as Samples of Icy Asteroids," *The Astrophysical Journal* 806(204), https://doi.org/10.1088/0004-637X/806/2/204.

[41] D. Jewitt, H. Hsieh, and J. Agarwal, 2015, "The Active Asteroids," Pp. 221–241 *Asteroids IV*, P. Michel, F.E. DeMeo, and W.F. Bottke, eds., Tucson: University of Arizona Press.

[42] F. Vilas and M.J. Gaffey, 1989, "Phyllosilicate Absorption Features in Main-Belt and Outer-Belt Asteroid Reflectance Spectra," *Science* 246:790–792, https://doi.org/10.1126/science.246.4931.790.

[43] A.S. Rivkin, B.E. Clark, M. Ockert-Bell, et al., 2011, "Asteroid 21 Lutetia at 3 μm: Observations with IRTF SpeX," *Icarus* 216:62–68, https://doi.org/10.1016/j.icarus.2011.08.009.

[44] A.S. Rivkin and J.P. Emery, 2010, "Detection of Ice and Organics on an Asteroidal Surface," *Nature* 464:1322–1323, https://doi.org/10.1038/nature09028.

[45] H. Campins, K. Hargrove, N. Pinilla-Alonso, et al., 2010, "Water Ice and Organics on the Surface of the Asteroid 24 Themis," *Nature* 464:1320–1321, https://doi.org/10.1038/nature09029.

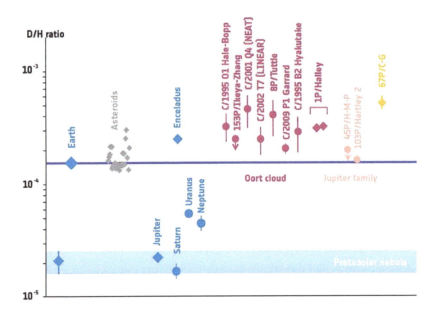

FIGURE 2-6 The different values of the deuterium-to-hydrogen ratio (D/H) in water observed in various bodies in the solar system.
SOURCE: K. Altwegg, M. Bauer, and M. Taylor, 2014, "Rosetta Fuels Debate on Origin of Earth's Oceans," downloaded from the ESA Image Archive.

buried ice on spherical bodies, within the top few meters of the surface, orbiting 2–3 AU from the Sun, can survive ~10^9 years.[46]

Organic- and volatile-rich asteroids provide fundamental information about the source of water and prebiotic compounds for the terrestrial planets.[47] As described in Box 2-2 and in Figure 2-6, spectroscopic measurements of the D/H ratios in cometary comae indicate that water ice in comets is more D-rich than the water at the surface of Earth, constraining the amount of volatile material that could be delivered from cometary impacts. Furthermore, dynamical simulations of the formation of terrestrial planets suggest that the outer asteroid belt was the primary source of impactors on the early Earth. The discovery of MBCs suggests that similar bodies may have delivered water and other volatiles to the inner solar system.[48]

SPACECRAFT VISITS

Relatively few small bodies have been visited by spacecraft (Table 2-1), given the large number of small bodies, which means that nearly all knowledge of small bodies throughout the solar system is based on Earth-based observations and models, along with meteorite and sample studies. All the JFCs that have been visited by spacecraft demonstrate remarkable diversity (e.g., Figure 2-2). A small number of spacecraft have visited comets in situ, including coma sample-return missions: after flying by 1P/Halley,

[46] N. Schorghofer, 2008, "The Lifetime of Ice on Main Belt Asteroids," *The Astrophysical Journal* 682:697–705, https://doi.org/10.1086/588633.

[47] D.S. Lauretta, O. Barnouin-Jha, M.A. Barucci, et al., 2009, "Astrobiology Research Priorities for Primitive Asteroids," Lunar and Planetary Institute, https://www.lpi.usra.edu/decadal/sbag/topical_wp/lauretta_etal.pdf.

[48] H.H. Hsieh and D. Jewitt, 2006, "A Population of Comets in the Main Asteroid Belt," *Science* 312:561–563, https://doi.org/10.1126/science.1125150.

TABLE 2-1 Small Bodies Visited by Spacecraft

Name	Body Type	Mission	Year
1P/Halley	Comet	Giotto, Vega 1, Vega 2	1986
951 Gaspra	S-type MBA	Galileo (flyby)	1991
26P/Grigg-Skjellerup	Comet	Giotto (no images)	1992
243 Ida	S-type MBA	Galileo (flyby)	1994
253 Mathilde	C-type MBA	NEAR Shoemaker (flyby)	1997
433 Eros	S-type NEO	NEAR Shoemaker (orbiter/lander)	1998–2001
2685 Masursky	S-type MBA	Cassini (distant flyby)	2000
19P/Borrelly	Comet	Deep Space 1	2001
5535 Annefrank	S-type MBA	Stardust (flyby)	2002
25143 Itokawa	S-type NEO	Hayabusa (sample return)	2003–2010
81P/Wild-2	Comet	Stardust (coma sample return)	2004–2006
9P/Tempel-1	Comet	Deep Impact with Impactor Stardust-NExT	2005, 2011
2867 Steins	E-type MBA	Rosetta (flyby)	2008
103P/Hartley-2	Comet	EPOXI (Deep Impact)	2010
21 Lutetia	M-type MBA	Rosetta (flyby)	2010
4 Vesta	V-type MBA	Dawn (orbiter)	2011–2012
4179 Toutatis	Stype NEO	Chang'e 2 (flyby)	2012
67P/Churyumov-Gerasimenko	Comet	Rosetta (orbiter and lander)	2014–2016
Pluto	KBO/dwarf planet	New Horizons (flyby)	2015
1 Ceres	C-type MBA/dwarf planet	Dawn (orbiter)	2015–2018
486958 Arrokoth	KBO	New Horizons (flyby)	2019
162173 Ryugu	C-Complex NEO	Hayabusa2 (sample return)	2019
101955 Bennu	B-type NEO	OSIRIS-REx (sample return)	2020
16 Psyche	M-type MBA	Psyche (orbiter)	Launch 2023 or 2024
52246 Donaldjohansen	C-type MBA	Lucy (flyby)	2025
3548 Eurybates	Jupiter Trojan	Lucy (flyby)	2027
15094 Polymele	Jupiter Trojan	Lucy (flyby)	2027
11351 Leucus	Jupiter Trojan	Lucy (flyby)	2028
21900 Orus	Jupiter Trojan	Lucy (flyby)	2028
617 Patroclus and Menoetius	Jupiter Trojan	Lucy (flyby)	2033

NOTE: KBO, Kuiper Belt object; MBA, Main Belt asteroid; NEO, near-Earth object.

Giotto flew-by comet 26P/Grigg-Skjellerup;[49] Deep Space One passed by 19P/Borrelly;[50] and 81P/Wild 2 was visited by the Stardust spacecraft before its extended mission, NExT, flew by 9P/Tempel 1.[51] Cometary dust, collected in the coma of 81P/Wild 2 was returned to Earth for in-depth analysis. Other comets visited were 9P/Tempel 1 by Deep Impact[52] and 103P/Hartley 2 by the (renamed) EPOXI spacecraft,[53] before ESA's Rosetta mission encountered comet 67P/C-G.[54] Rosetta was the first spacecraft to rendezvous with a comet, and it orbited the nucleus for about 2 years with the Philae lander module being deployed onto the comet's nucleus early in the orbiting phase.

When the Galileo spacecraft flew by (243) Ida in 1994, the remarkable discovery of a natural satellite at Ida (Dactyl) was made, not observable from Earth.[55] Galileo cameras also provided critical insights to space weathering processes on S-type asteroids, by noting the spectrally blue nature of craters on Ida, compared to the relatively spectrally reddish overall nature of the space-weathered surface.[56] The visits by the Hayabusa2 and OSIRIS-REx spacecraft to Ryugu and Bennu, respectively, have provided insights into the nature and evolution of these "spinning top" shaped asteroids,[57] along with the remarkable particle ejection processes seen at Bennu.[58] Almost every small body studied close-up by spacecraft has had unique characteristics, sometimes quite unexpected.

LINKS TO THE EARLY SOLAR SYSTEM

Dynamic modeling combined with observational constraints indicates that the rocky bodies in the inner Main Belt likely formed somewhere close to their current locations and have been thermally metamorphosed or melted. Models such as the Grand Tack/Nice models suggest that planetary migration of Jupiter and Saturn produced sweeping resonance through the main asteroid belt and dislodged most of the asteroids. The resulting liberated asteroids could have been responsible for the impact cataclysms that occurred on all terrestrial planets and satellites around 4 billion years ago. Some of the low-albedo objects in the Main Belt likely formed in the outer solar system and ended up in the Main Belt as a result of giant planet migration. Some ended up in cold enough locations such that pre-existing ice never melted, and some were aqueously altered due to melting of ice (Figure 2-7). The role of ^{26}Al in heating and melting is

[49] M.G. Grensemann and G. Schwehm, 1993, "Giotto's Second Encounter: The Mission to Comet P/Grigg-Skjellerup," *Journal of Geophysical Research* 98:20907–20910, https://doi.org/10.1029/93JA02528.

[50] L.A. Soderblom, T.L. Becker, G. Bennett, et al., 2002, "Observations of Comet 19P/Borrelly by the Miniature Integrated Camera and Spectrometer Aboard Deep Space 1," *Science* 296:1087–1091, https://doi.org/10.1126/science.1069527.

[51] D. Brownlee, P. Tsou, J. Aléon, et al., 2006, "Comet 81P/Wild 2 Under a Microscope," *Science* 314(1711), https://doi.org/10.1126/science.1135840.

[52] M.F. A'Hearn, M.J.S. Belton, W.A. Delamere, et al., 2005, "Deep Impact: Excavating Comet Tempel 1," *Science* 310:258–264, https://doi.org/10.1126/science.1118923.

[53] M.F. A'Hearn, M.J.S. Belton, W.A. Delamere, et al., 2011, "EPOXI at Comet Hartley 2," *Science* 332(1396), https://doi.org/10.1126/science.1204054.

[54] K.-H. Glassmeier, H. Boehnhardt, D. Koschny, E. Kührt, and I. Richter, 2007, "The Rosetta Mission: Flying Towards the Origin of the Solar System," *Space Science Reviews* 128:1–21, https://doi.org/10.1007/s11214-006-9140-8.

[55] M.J.S. Belton, C.R. Chapman, P.C. Thomas, et al., 1995, "Bulk Density of Asteroid 243 Ida from the Orbit of Its Satellite Dactyl," *Nature* 374(6525):785–788, https://doi.org/10.1038/374785a0.

[56] C. Chapman, 1996, "S-Type Asteroids, Ordinary Chondrites and Space Weathering: The Evidence from Galileo's Flybys of Gaspra and Ida," *Meteoritics and Planetary Science* 31:699–725, https://doi.org/10.1111/j.1945-5100.1996.tb02107.x.

[57] P. Michel, R.-L. Ballouz, O.S. Barnouin, et al., "Collisional Formation of Top-Shaped Asteroids and Implications for the Origins of Ryugu and Bennu," *Nature Communications* 11(1), https://doi.org/10.1038/s41467-020-16433-z.

[58] D.S. Lauretta, C.W. Hergenrother, S.R. Chesley, et al., 2019, "Episodes of Particle Ejection from the Surface of Active Asteroid (101955) Bennu," *Science* 366(6470), https://doi.org/10.1126/science.aay3544.

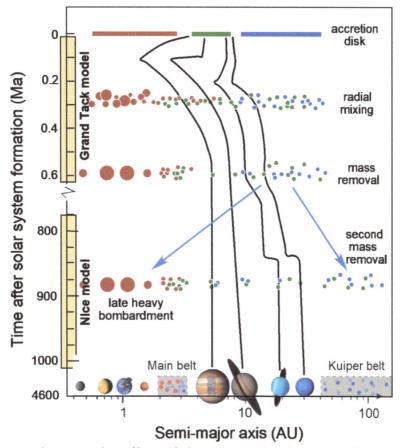

FIGURE 2-7 Cartoon demonstrating effects of giant planet migration on the distribution of small bodies in the solar system.
SOURCE: F.E. DeMeo and B. Carry, 2014, "Solar System Evolution from Compositional Mapping of the Asteroid Belt," *Nature* 505:629–634, https://doi.org/10.1038/nature12908. Reprinted by permission from Springer Nature, Copyright 2014.

also important; one model[59] suggests that bodies forming closer to the Sun accreted earlier and thus had more ^{26}Al, which would have led to silicate melting, thermal metamorphism, and ice melting. From a planetary protection perspective, the most astrobiologically relevant targets are volatile-rich carbonaceous asteroids. But, with the exception of Ceres, aqueous fluids only existed ~4.5 billion years ago.

Understanding the origin of organic compounds in early solar system materials is central to astrobiology. Individual asteroids are "astrobiological time capsules"[60] that preserve a record of the evolution of volatiles and organics starting in the interstellar medium, through the birth and early evolution of the solar system, to present-day space weathering at asteroid surfaces.

Small bodies record the radial compositional gradients of material that were present in the protosolar disk, and they represent all stages of the formation and early evolution of the solar system. Primitive small bodies are the debris left over from planet formation and they contain examples of the primordial ingredients from which the planets and life arose. Small bodies record internal processing such as aqueous

[59] R.E. Grimm and H.Y. McSween, 1993, "Heliocentric Zoning of the Asteroid Belt by Aluminum-26 Heating," *Science* 259(5095):653–655, https://doi.org/10.1126/science.259.5095.653.

[60] D.S. Lauretta, O. Barnouin-Jha, M.A. Barucci, et al., 2009, "Astrobiology Research Priorities for Primitive Asteroids," Lunar and Planetary Institute, https://www.lpi.usra.edu/decadal/sbag/topical_wp/lauretta_etal.pdf.

alteration, thermal metamorphism, melting and differentiation. These evolved small bodies record processes that occurred during the formation and evolution of planets. Small bodies thus trace growth from primordial condensates and presolar/interstellar grains to pebbles, planetesimals, and planetary embryos, to planets. Because catastrophic collisions can completely shatter protoplanets, small bodies can be samples of core, mantle, or crustal material of once much larger bodies. Indeed, this is the only known reservoir of accessible core and, potentially, deep mantle material. Small bodies also record the history of the solar system such as the dynamical evolution of the solar system, the evolution of surface materials through time and as they approach the Sun, and the primordial cosmo-chemical gradients established within the solar nebula.

> **Finding 1:** The primary astrobiological value of small solar system bodies is that some of these bodies contain prebiotic organic compounds that are relevant to the study of the origin of life in the solar system.

As a consequence, most small-body missions are either planetary protection Category I or II and thus do not require provisions to ensure spacecraft cleanliness. There are a few exceptions, most notably Ceres (see Box 2-1), and these may require special attention as will be discussed in Chapter 3.

CONDITIONS ON SMALL BODIES WITH RESPECT TO SURVIVAL AND PROLIFERATION

In the framework of planetary protection for solar system small bodies, scientific understanding of life is based on our knowledge of life on Earth. Because of the incomplete current state of knowledge about small bodies as well as the survival-limits of life, all judgments regarding biological potential are qualitative, not quantitative. On Earth, liquid water is the solvent for life, and desiccation alone will prevent cell proliferation, but can allow for survival of dormant cells in which intracellular water is lost and metabolism is undetectable, yet can resume activity upon aquation. As examples, archaea, bacteria, and fungi possess a number of strategies that allow them to survive desiccation in the form of spores or when simply dried as vegetative cells.[61,62,63] Of course, small bodies exist in the vacuum of outer space, and any foreign contaminating cells would be frozen, desiccated, and unable to repair genetic damage caused by unremitting solar and galactic cosmic radiation (GCR). In determining the likelihood of forward contamination, considerations of water content and radiation on small bodies thus allow for a conservative approach toward gauging survival of life on asteroids and comets.

The survivability of frozen and desiccated microorganisms transported to surface environments of small bodies would be governed by radiation exposure, which varies across the surface and subsurface of small bodies and with distance from the Sun. On an asteroid's surface, ultraviolet C (UVC, light in the 200–280 nm range) will inactivate cells within days to months.[64,65] For microorganisms shielded from UVC and heat in the near subsurface (0.1 m), ionizing radiation, largely due to GCR and solar protons,

---

[61] K.L. Anderson, E.E. Apolinario, and K.R. Sowers, 2012, "Desiccation as a Long-Term Survival Mechanism for the Archaeon Methanosarcina Barkeri," *Applied and Environmental Microbiology* 78(5):1473–1479, https://doi.org/10.1128/AEM.06964-11.

[62] P. Setlow, 2007, "I Will Survive: DNA Protection in Bacterial Spores," *Trends in Microbiology* 15(4):172–180, https://doi.org/10.1016/j.tim.2007.02.004.

[63] V. Mattimore and J.R. Battista, 1996, "Radioresistance of Deinococcus Radiodurans: Functions Necessary to Survive Ionizing Radiation Are Also Necessary to Survive Prolonged Desiccation," *Journal of Bacteriology* 178(3):633–637, https://doi.org/10.1128/jb.178.3.633-637.1996.

[64] NASEM, 2021, *Report Series: Committee on Planetary Protection: Evaluation of Bioburden Requirements for Mars Missions*, Washington, DC: The National Academies Press, https://doi.org/10.17226/26336.

[65] A. Vicente-Retortillo, G.M. Martínez, N.O. Rennó, M.T. Lemmon, M. de la Torre-Juárez, and J. Gómez-Elvira, 2020, "In Situ UV Measurements by MSL/REMS: Dust Deposition and Angular Response Corrections," *Space Science Reviews* 216:97, https://doi.org/10.1007/s11214-020-00722-6.

would extinguish survival over thousands of years (~0.1 Gy/yr); but for desiccated and frozen microbes transported to the deep subsurface (10 m), only internal background radiation is expected to limit survival over millions of years, as on Earth (~0.5 mGy/yr).[66] No dormant DNA-based lifeforms would be expected to withstand billions of years of continuous background radiation on asteroids because, without growth and DNA repair, contaminating cells would inevitably be destroyed.[67] With regard to planetary protection for missions to asteroids, therefore, the committee considerations support the conclusion that any forward biological contamination will be harmless because cells would not be able to grow or proliferate and thus would be sterilized over time, leaving only macromolecular cellular debris.

No known small bodies have atmospheres (with the exception of cometary comae), and with the possible exceptions of Ceres, Pluto, and Themis, there is no known geologic activity on any small bodies, limiting the transport of material across their surfaces. An exception might be the particle ejection seen to occur on Bennu (and likely on other bodies, probably the result of thermal processing); another possible exception is electrostatic levitation[68] or redistribution of surface material (e.g., on top-shaped asteroids) via rotational forces. Moreover, comets have extensive resurfacing due to repeated passes in the inner solar system, as described in the next chapter. The lack of atmosphere also means that temperatures are extreme and there is no protection from radiation, nor is there liquid water. Implications for planetary protection are that any terrestrial microbes are extremely unlikely to survive, much less proliferate, on any small body.

The absence of geologic and weather processes both on asteroids and comets further means that their bulk materials and composition have not changed significantly since they were formed in the early solar system, 4.6 billion years ago. Indeed, the abundances of asteroidal and cometary parent molecules and those observed in the interstellar medium show a striking similarity for many of the simple chemical species. Recent discoveries, though, have changed the perception that asteroids and comets contain only simple volatile molecules like CO, CO_2, NH_3, and water, to include multiple amino acids.[69] So, small-body objects represent targets for the study of materials which not only gave rise to planets but also to the organic precursors of life on Earth. Although small bodies may have served as a delivery mechanism for the building blocks of life, the small-body objects themselves are not relevant to prebiotic biochemical evolution of macromolecular life as occurred on Earth.

These findings are exemplified by recent small-body missions: Stardust captured organic matter samples from Comet 81P/Wild2.[70] Rosetta's close proximity to the coma of comet 67P/C-G allowed it to detect numerous organic species including aromatic hydrocarbons, oxygenated hydrocarbons, and a diverse population of sulfur-bearing molecules in addition to many inorganic species.[71] Similarly, the Hayabusa2 sample-return mission showed surface-excavated materials of the rubble-pile asteroid Ryugu

[66] See Chapter 5 and Appendix A in NASEM, 2002, *The Quarantine and Certification of Martian Samples*, Washington, DC: The National Academies Press, https://doi.org/10.17226/10138.

[67] D. Ghosal, M.V. Omelchenko, E.K. Gaidamakova, et al., 2005, "How Radiation Kills Cells: Survival of Deinococcus Radiodurans and Shewanella Oneidensis Under Oxidative Stress," *FEMS Microbial Reviews* 29(2):361–375, https://doi.org/10.1016/j.femrre.2004.12.007.

[68] M.S. Robinson, P.C. Thomas, J. Veverka, S. Murchie, and B. Carcich, 2001, "The Nature of Ponded Deposits on Eros," *Nature* 413:396–400, https://doi.org/10.1038/35096518.

[69] T. Yada, M. Abe, A. Nakato, et al., 2022, "Preliminary Analyses on Bulk and Individual Ryugu Samples Returned by Hayabusa2," *LPI Contributions* 2678(1831), 53rd Lunar and Planetary Science Conference, https://www.hou.usra.edu/meetings/lpsc2022/pdf/1831.pdf.

[70] S.A. Sandford, J. Aléon, C.M.O'D. Alexander, et al., 2006, "Organics Captured from Comet 81P/Wild2 by the Stardust Spacecraft," *Science* 330:468–472, https://doi.org/10.1126/science.1135841.

[71] M. Schuhmann, K. Altwegg, H. Balsiger, et al., 2019, "Aliphatic and Aromatic Hydrocarbons in Comet 67P/Churyumov-Gerasimenko Seen by ROSINA," *Astronomy and Astrophysics* 630(A31), https://doi.org/10.1051/0004-6361/201834666.

to contain more than 10 types of amino acids including glycine and L-alanine.[72] In summary, with regard to planetary protection of asteroids and comets, the committee concludes that forward contamination by microorganisms would not be harmful because cells would not proliferate in vacuo.[73]

> **Finding 2:** Based on current knowledge, it is highly improbable that small bodies harbor extinct or extant life, or that terrestrial microbes carried by a landing spacecraft can proliferate on a small body. Furthermore, given the short timescales of inactivation by ultraviolet C (UVC, or short-wavelength ultraviolet light, from 200–280 nm) radiation, there is no realistic likelihood that terrestrial microbes delivered by a spacecraft to a particular small body can be transported to another small body in a timeframe comparable to the timescales relevant for missions to small bodies (i.e., contaminating body A will not threaten body B).

---

[72] T. Yada, M. Abe, A. Nakato, et al., 2022, "Preliminary Analyses on Bulk and Individual Ryugu Samples Returned by Hayabusa2," *LPI Contributions* 2678(1831), 53rd Lunar and Planetary Science Conference, https://www.hou.usra.edu/meetings/lpsc2022/pdf/1831.pdf.

[73] With regard to planetary protection, the great radiation-survivability of desiccated *D. radiodurans* cells support two findings. First, it must be assumed that any forward contamination of small bodies with terrestrial microorganisms would essentially be permanent, over mission timeframes of thousands of years, and this could complicate scientific efforts in the search for life, even though terrestrial microbes would not proliferate in vacuo. Second, if whole viable *D. radiodurans* cells could survive the equivalent of 1.4 Myr in the near subsurface environments of asteroids, then their macromolecules will survive much, much longer.

3

Criteria for Planetary Protection Categorization of Small Body Missions

Small bodies in the solar system have preserved information from different eras of the evolution of our solar system, from the composition of the presolar nebula that is preserved in cometary nuclei to the more evolved compositions found in some asteroids. As shown in Chapter 2, the small bodies include a wide variety of different asteroid types, the comets, Trojans, Centaurs, and Kuiper Belt objects (KBOs). Box 2-1 discusses the specific case of Ceres.

CRITERIA CONSIDERED FOR CATEGORIZATION

In considering planetary protection categorization for missions to the different types of small bodies, in particular between Category I and Category II, the committee considered the following criteria.[1]

Size of Population

The size of the population of any one subclass of bodies could be considered as a criterion for determining a mission's planetary protection categorization. If a group of some type of small body is sufficiently large, one could postulate that missions to some percentage of that type of small body could possibly be allowed to be categorized as Category I given that there are plenty of other bodies of that group that could remain at Category II. However, the committee concluded that not enough is known about any one subclass to use the size of a population as a criterion for planetary protection categorization.

Status of Knowledge

The status of knowledge of any subclasses of small bodies might, in theory, be used to argue that missions to some small bodies can be classed as Category I rather than Category II because they are well-understood and therefore it might be considered acceptable to allow a mission to visit with no requirements for planetary protection documentation (Category I).

The issue here is that current knowledge of small bodies in the solar system is limited. Although some small bodies have been imaged with high resolution radar or adaptive optics data, currently most information about small solar system bodies is derived from ground-based observations that show the objects merely as points of light. Very little is known about Centaurs, Trojans, and even outer Main Belt asteroids (MBAs); these types of bodies never been visited. At the same time, each mission that has visited a small body has found it to be somehow unique and scientifically important; furthermore, taxonomic classes cannot be used with certainty to understand individual targets using Earth-based information alone. Thus, too little is known about any one class of small bodies to use current knowledge as the only criterion for planetary protection mission categorization. However, the committee notes that it

---

[1] Simultaneously, the committee considered the subtle differences between Category I and Category II, as discussed in Chapter 4.

is important that the status of knowledge of small bodies be considered periodically and that planetary protection categorization be reconsidered as needed. For instance, nearly all missions to small bodies are currently classified as Category I or Category II. Some future observations of some body or class of bodies could make missions to those bodies appropriately classed as Category IV; for example, information from measurements on the presence of geologic activity. Alternately, future observations of some type of small body could indicate that follow-on missions to those types of bodies could be changed from Category II to Category I.

Likelihood of Revisiting

The foundational concept of planetary protection is the preservation of the scientific integrity of a body or region for future astrobiological studies of that body or region. If there is no likelihood that a target will be revisited for astrobiological study, there is no compelling reason to not categorize a mission to that target as Category I. However, the committee concluded, along the lines of the above criterion on status of knowledge, that one cannot assume, prior to ever visiting a target, that there will be no scientific reason to revisit it.

Geological Activity and Resurfacing

Currently, with some possible exceptions, the only known Main Belt small body demonstrating evidence of potential geologic activity (based on a very limited suite of spacecraft visits) is Ceres. Ceres is special because of the Dawn discovery of organics on the surface of Ceres, and because Dawn found Ceres to be a water-rich body with evidence of brine-driven cryovolcanism occurring very recently in history and perhaps even currently (see Box 2-1). Most of Ceres's surface is more than a few hundred million years old, and a large fraction of it is >1 Gyr,[2] based on crater counts. Occator crater, however, has a fresh morphology, and could experience transfer of material from the deep interior via impact-produced fractures, currently or in the recent past. More robust evidence for ongoing material exposure[3] is in the form of hydrated salt (hydrohalite, $NaCl \cdot 2H_2O$) found at the top of a dome in the center of one of Occator's faculae.[4] Because hydrated salts are not stable in vacuum at the temperature on Ceres's surface and dehydrate over a short timescale (months or years),[5] this material is likely currently being emplaced.

Comets are active planetary bodies because they display evidence for nearly all fundamental geological processes, which include impact cratering, tectonism, and erosion. Comets also display sublimation-driven outgassing, which is comparable to volcanism on larger planetary bodies in that it provides a conduit for delivering materials from the interior to the surface. However, in the domain of active geological bodies, comets occupy a special niche because their geologic activity is almost exclusively driven by externally supplied energy (i.e., solar energy) as opposed to an internal heat source, which makes them "seasonally active" bodies. During their active phase approaching the Sun, comets also develop a transient atmosphere that interacts with the surface and contributes to its evolution, particularly

---

[2] S.C. Mest, D.C. Berman, D.L. Buczkowski, et al., 2021, "The Global LAMO-Based Geologic Map of Ceres," 2021 Annual Meeting of Planetary Geologic Mappers, 2610(7035), https://www.hou.usra.edu/meetings/pgm2021/eposter/7035.pdf.

[3] J.C. Castillo-Rogez, M. Neveu, V. Vinogradoff, et al., 2022, "Science Drivers for the Future Exploration of Ceres: From Solar System Evolution to Ocean World Science," *The Planetary Science Journal* 3(64), https://doi.org/10.3847/PSJ/ac502b.

[4] M.C. De Sanctis, E. Ammannito, A. Raponi, et al., 2020, "Fresh Emplacement of Hydrated Sodium Chloride on Ceres from Ascending Salty Fluids," *Nature Astronomy* 4:786–793, https://doi.org/10.1038/s41550-020-1138-8.

[5] C. Bu, G. Rodriguez Lopez, C.A. Dukes, L.A. McFadden, J.-Y. Li, and O. Ruesch, 2017, "Instability of Magnesium Sulfate Hexahydrate (MgSO4.6H2O) on Ceres: Laboratory Measurements," *LPI Contributions* 2996, 48th Annual Lunar and Planetary Science Conference, https://www.hou.usra.edu/meetings/lpsc2017/pdf/2996.pdf.

by transporting materials across the surface. Variations in solar energy input on diurnal and seasonal scale cause buildup of thermal stresses within consolidated materials that lead to weathering through fracturing, and eventually mass-wasting. The commonly irregular shapes of comets also play a major role in their evolution by leading to

1. Non-uniform gravitational forces that affect material movement across the surface; and
2. Spatially heterogeneous outgassing patterns that affect the comet's orbital dynamics and lead to tidal stresses that can further fracture the nucleus.[6]

The literature discusses the seasonal mass transfer on the nucleus of 67P/C-G, which shows the complexity of the problem. To date, data taken by Rosetta provide the only set of observations of a comet nucleus along its orbit with the onset of activity, through perihelion passage until the next "dormant" phase at large heliocentric distance. An average erosion of the surface of about 0.4 m was determined,[7] depending on the model used; however, this value for surface erosion can reach even 20 m during a single perihelion passage due to the strong insolation around the comet's southern summer solstice. Taking a surface erosion or ablation of about 1 m for a comet nucleus as an average per orbit around the Sun is a conservative estimate. This means that the surface of a comet nucleus changes continuously along its orbit when it is inside 3 AU and a revisiting spacecraft will thus see a totally different surface morphology, and perhaps composition.

For some geologically active bodies, resurfacing can imply a "freshening" of the surface between spacecraft visits. This is likely the case for comets and could be the case for some active asteroids. One can consider that a body or region that is frequently resurfaced could be a prime candidate for Category I missions, because there may be no need for contamination control. Alternately, such resurfacing could, perhaps, be a mechanism for burial of terrestrial microbes, possibly to a subsurface location where they can survive.

Size of Target Body

Larger asteroids ($\gtrsim$100 km diameter) are generally considered to be primordial whereas smaller bodies are likely collisional fragments, as discussed in Chapter 2. Thus, in investigations of original, unaltered primitive material that is left over from the early solar system, these larger small bodies are more likely to hold unprocessed or unaltered evidence of the early solar system environment. Larger targets are thus considered to hold more astrobiological interest for potential future science missions than small fragmentary bodies.

Composition

As noted in Finding 1 of this report, small bodies throughout the solar system are primarily astrobiologically important not because they could harbor life but because they hold compositional clues to the early solar system and to the distribution of organics and other volatiles that are the building blocks of life and that could have brought water and life to Earth. Consequently, the committee considers those small bodies that are volatile-rich and/or organic-rich to be more astrobiologically important than the rocky types. The volatile- and/or organic-rich small bodies are generally of the C-complex group (C-,

---

[6] M.R. El-Maarry, O. Groussin, H.U. Keller, et al., 2019, "Surface Morphology of Comets and Associated Evolutionary Processes: A Review of Rosetta's Observations of 67P/Churyumov-Gerasimenko," *Space Science Reviews* 215(36), https://doi.org/10.1007/s11214-019-0602-1.

[7] H.U. Keller, S. Mottola, S.F. Hviid, et al., 2017, "Seasonal Mass Transfer on the Nucleus of Comet 67P/Chuyumov-Gerasimenko," *Monthly Notices of the Royal Astronomical Society* 469:S357–S371, https://doi.org/10.1093/mnras/stx1726.

Cb-, Ch-, Cg-, Cgh-, and B-types) or the P- and D- types, primarily with semi-major axes of >2.5 AU (Figure 2-5), while the rocky types are generally S-types; numerous other taxonomic types have been identified, including the metallic M-types. Comets are likely even more primordial than asteroids, given their origin in the Kuiper Belt or Oort Cloud; thus, despite (or because of) their active resurfacing, the primitive nature of comets makes them astrobiologically important. Ceres has been found to have local surface deposits of organic matter,[8] adding to the astrobiological interest of this special world.

The committee concludes that target composition (namely, taxonomic type for asteroids) is the primary factor to use in categorization decisions. The committee does not consider Trans-Neptunian objects (TNOs) beyond 50 AU in this assessment. Furthermore, the committee concludes that planetary protection assessment for missions to Pluto and other KBOs dwarf planets will no doubt need to be considered once such a mission is prioritized; the committee anticipates that any such mission will be a scientific mission and thus scientific requirements will likely outweigh planetary protection cleanliness requirements.

> **Finding 3:** The committee does not find a need to change current categorization of missions to small bodies. Category II is an appropriate planetary protection category for missions to relatively primitive, volatile-rich, and organic-bearing small bodies that have astrobiological importance—including C-complex (C-, Cb-, Ch-, Cg-, Cgh-, and B-types), P-type, and D-type MBAs and near-Earth objects (NEOs), Trojans, comets, KBOs, and Centaurs. These objects have the potential to provide insights about prebiotic chemistry. Category II requires the provision of information that is important for future missions to the same targets, such as spacecraft impact or landing sites. The chemistry of other small bodies is likely not of astrobiological interest, and Category I is an appropriate category for missions to these objects, including rocky, metamorphosed, and metallic NEOs and MBAs.

> **Finding 4:** Current scientific knowledge regarding some large asteroids (e.g., low-albedo objects $\gtrsim$100 km in diameter and having an orbital semi-major axis greater than ~2.5 AU) is not sufficient to support well-informed categorization of missions to those objects, but Category II is acceptable until future reassessment. Ceres is a notable example of a large object with recently discovered importance to astrobiology and thus future missions to Ceres merit reassessing in terms of planetary protection categorization. Future missions to Ceres will likely require more rigorous planetary protection protocols than Category II.

> **Finding 5:** The committee endorses the periodic reassessment of the planetary protection categorization scheme for all small bodies on a regular cadence, allowing for the most recent science information to be taken into account.

---

[8] M.C. De Sanctis, E. Ammannito, H.Y. McSween, et al., 2017, "Localized Aliphatic Organic Material on the Surface of Ceres," *Science* 355:719–722, https://doi.org/10.1126/science.aaj2305.

4

Implications of Planetary Protection Category I Versus Category II for Small Body Missions

Both National Aeronautics and Space Administration (NASA) and Committee on Space Research (COSPAR) policies categorize all missions to small solar system bodies as either Category I or Category II. Current NASA Policy (NASA Procedural Requirement [NPR] 8715.24) includes the following description of planetary target priorities to differentiate between Category I and Category II missions:

Category I: Not of direct interest for understanding the process of chemical evolution or where exploration will not be jeopardized by terrestrial contamination.

Category II: Of significant interest relative to the process of chemical evolution but only a remote chance that contamination by spacecraft could compromise future investigation.

Using these broad definitions, both NASA and COSPAR currently consider missions to undifferentiated, metamorphosed asteroids as Category I and missions to comets and all other types of asteroids as Category II. No cleanliness requirements or organics inventories[1] are imposed on either Category I or Category II missions. Category I missions carry no planetary protection requirements at all. The planetary protection requirements for Category II missions are limited to relatively routine information about the mission. This includes documenting the intended target body, the mission intent (flyby, landing, or impact), the planned trajectory, the post launch trajectory status, and the final disposition of the spacecraft.

For some small-body missions, if the trajectory is intended to fly by, or has a significantly high risk of approaching a Category III/IV target body (such as Mars, Europa, or Enceladus), then the mission will receive a Category III categorization and additional planetary protection requirements will apply, such as those for spacecraft cleanliness. NASA missions to small bodies that received Category III categorization include Dawn and Psyche, both due to the intent to fly past Mars.

The Juno mission to Jupiter provides an example of mission receiving a revised categorization as the result of its extended mission plan. The initial Juno mission was Category II. After launch, the mission was extended to include flybys of Europa, Ganymede, and Io, and received a revised Category III categorization.

PLANETARY PROTECTION DOCUMENTATION FOR CATEGORY II MISSIONS

NPR 8715.24 lists six possible document requirements for Category II missions:[2]

- Planetary Protection Requirements Document
- Planetary Protection Implementation Plan

---

[1] The requirement for an organics inventory may be required for certain missions to the Earth's Moon; such missions carry a Category IIb categorization. Category IIb is not associated with any missions to small bodies.

[2] NASA, 2021, "NASA Procedural Requirements: Biological Planetary Protection for Human Missions to Mars," NPR 8715.24, Washington, DC: NASA, September 24.

- Pre-Launch Planetary Protection Report
- Post-Launch Planetary Protection Report
- Extended Mission Planetary Protection Report
- End-of-Mission Planetary Protection Report

The document further notes that "Some missions may be able to demonstrate compliance with planetary protection requirements with a reduced document set."[3]

The planetary protection categorization of NASA missions is guided by COSPAR policy and guidelines. COSPAR Policy provides additional detail (see Appendix B) on four specific documentation requirements for Category II missions:

1. A brief Planetary Protection Plan outlining intended or potential impact targets,
2. A brief Pre-Launch Planetary Protection Report detailing impact avoidance strategies, if required,
3. A brief Post-Launch Planetary Protection Report detailing actual trajectory and any updates of previous documentation, and
4. An End-of-Mission Report providing the final actual disposition of the launched hardware and impact location.

Documents 2, 3, and 4 can be viewed as a progression. Prior to launch, Document 2 captures all the essential information required of the mission for planetary protection purposes. Document 3 can either supersede or supplement that information, including post-launch trajectory data. Likewise, Document 4 either supersedes or supplements Documents 2 and 3 with summary updates at the conclusion of the mission. If the intent of a mission is to fly by or orbit the target body, but not land on the target body, then the information requested will include a discussion of the approach to avoiding unintentional impact with the target or any other Category II object. For some Category II missions, the NASA Office of Planetary Protection has required additional trajectory analysis to demonstrate a sufficiently low risk of impact (and potential contamination) of other solar system bodies, particularly Mars and Europa.

Past practice within NASA has been that the specifics of the documentation required for a Category II mission is detailed in a Categorization Letter issued by the Office of Planetary Protection.

Finding 6: Under current NASA and COSPAR planetary protection guidelines, Category II missions require only a minimal level of documented information, primarily target and impact/landing site.

ACCESS TO PLANETARY PROTECTION DOCUMENTATION

Planners for future small-body missions intending to study the processes of chemical evolution and the origin of life need to know what target bodies have been visited and whether any target body, as the result of a spacecraft landing or impact, may have been contaminated by material brought from Earth that would compromise their planned observations. There may be an interest in avoiding any previously visited target or landing site to ensure pristine scientific measurements. Alternatively, there may be benefits to returning to a previously visited target with the objective of continuing studies and/or operations, or minimizing mission costs by returning to a known environment using heritage spacecraft systems. In either case, knowledge of previous mission impact targets and locations, intentional or inadvertent, is important. The previous section of this chapter describes the relevant documentation to provide this information.

COSPAR recommends and NASA currently requires this information for Category II missions. Yet there does not appear to be a central repository to archive and retain the data for either current or future

---

[3] NASA, 2021, "NASA Procedural Requirements: Biological Planetary Protection for Human Missions to Mars," NPR 8715.24.3.1.1.6, Washington, DC: NASA, September 24.

needs. Given the increasing number of small-body missions (Table 2-1), such an archive is quickly becoming important. Moreover, past missions have been sponsored by large space-experienced government agencies that have their own archival requirements as well as collaborative agreements to facilitate information sharing. The future, however, is likely to bring many more small-body missions, including those initiated by new entrants to space exploration, such as private-sector entities with unknown archival standards. These changes argue for a cooperative archival effort, which could be accomplished by the establishment of a common repository.

Ideally, such a planetary protection document repository would include:

- Planetary protection documents from all space missions (international and domestic, government-sponsored, and private-sector), including the location of organic inventories if such inventories were required.
- Established standards for data storage.
- Established processes for archival, retrieval, and maintenance.
- International access with provisions to ensure the protection of proprietary and sensitive data.
- Appropriate authority to establish, maintain, and operate the archive.
- Agreement for sustained funding to maintain and upgrade the archive as needed in perpetuity.

NASA's Planetary Data System (PDS) sets an example of how such a repository might be designed and executed. Adding a new node to the PDS to accommodate planetary protection documents, might provide an efficient and cost-effective option to fulfil this archival need.

Such a planetary protection repository would benefit all missions interested in understanding the process of chemical evolution or the origin of life, regardless of their planetary protection category.

Finding 7: Access to information prepared in response to planetary protection requirements is important for planning future missions to certain small bodies to study chemical evolution and the origin of life. The committee was unable to confirm that an archive of planetary protection information currently exists.

Planetary Protection, Small Solar System Bodies, and Commercial Space Activities

COMMERCIAL SPACE INTEREST IN SMALL SOLAR SYSTEM BODIES

The concept of mining asteroids to extract their valuable elements and provide Earth with resources, in a future where terrestrial resources are depleted or extraction of them is not economically viable, has been around for more than 100 years.[1] Following these objectives, a few companies in the early 2010s announced their intentions to extract valuable metals from asteroids to bring them back to Earth for commercial purposes. While these early proposals have not been realized, renewed interest in mining small bodies by the private sector has lately been mostly focused on the extraction of basic elements such as water and carbon for production of propellants in space.

Water and carbon dioxide could be cold trapped after being extracted using a variety of heating techniques. After a separation and refining process, water could be then directly used for steam propulsion, heated to high temperatures to produce a plasma for electric propulsion, electrolyzed to obtain hydrogen and oxygen as propellants, or reacted with carbon dioxide to produce methane fuel. Thus, while a few companies are still interested in the extraction of metals to be sold on Earth (such as platinum group metals from M-type asteroids or iron and nickel from S-type bodies), most business plans focus on the use of asteroidal resources in space to fuel spacecraft bound for various destinations, rather than following the costly, energy-intensive practice of bringing all propellant from Earth. These plans include transport of extraterrestrial materials to Earth orbit (or cis-lunar space) rather than to the surface of Earth.

Based on this living-off-the-land approach, the main targets of many asteroid mining companies are C-complex near-Earth objects (NEOs) because of their carbon content and water bound in hydrated minerals (up to 20 percent in some spectral types). Other advantages include their close proximity (as compared to more numerous, but distant Main Belt asteroids), low in-flight propulsion requirements to access them (many with lower trajectory adjustment requirements than the Moon), and the better probability to detect them and evaluate their resource potential with ground or orbital observatories.

Extracting these valuable volatile elements may require accessing and processing the regolith or rocky surface by docking and anchoring spacecraft to the NEO. Alternatively, to avoid the complexity of landing on an ultra-low-gravity body, remote heating techniques using focused solar radiation to drill, extract, and trap volatiles have been proposed. In these techniques, concentrated solar energy thermally fractures the rock, constantly exposing new material, while fracturing of the body into finer particles is also caused by spalling, as water ice and other volatiles sublimate during the heating process. Cold traps or bags would be used to collect the extracted volatiles. These elements would then be processed in situ or taken to orbital depots to be refined and stored as propellants.

In terms of planetary protection implications, given the potential of finding prebiotic organic compounds on C-complex small bodies, Category II would apply to missions intended to access and remove volatile material from them (see Finding 3). For missions aimed at accessing and removing metals, volatiles, and other materials from S- or M-type bodies, Category I would apply.

---

[1] K. Tsiolkovsky, 1961, "Beyond the Planet Earth," *The Journal of the Royal Aeronautical Society* 65(612):846, https://doi.org/10.1017/S0368393100076100. Translated from Russian by Kenneth Syers.

COMMERCIAL SECTOR AWARENESS OF, AND COMPLIANCE WITH, PLANETARY PROTECTION REQUIREMENTS

Past reports on planetary protection policies identified persistent misperceptions, confusion, and uncertainty among actors in the commercial space industry about planetary protection requirements for missions to different solar system bodies.[2,3,4] In conducting this study on small bodies, the committee encountered misperceptions that planetary protection requirements for missions to small solar system bodies are more stringent and burdensome than, in fact, they are. This problem might arise because media coverage of planetary protection predominantly focuses on missions to Mars for which planetary protection requirements are more extensive.

However, as discussed above, commercial sector interest in small solar systems bodies is, for the foreseeable future, focused on NEOs that might contain exploitable water or mineral resources. National Aeronautics and Space Administration (NASA) and Committee on Space Research (COSPAR) policies currently categorize missions to such asteroids as Category II and require the provision of information as outlined in Chapter 4. Category II does not impose any requirements for spacecraft cleanliness, information about organics carried by spacecraft, or data about what the mission does or discovers on the small body.

The committee also found that there seem to be misperceptions in the science community about current cleanliness requirements for missions to small bodies, with the misunderstanding (among many) that current planetary protection policies for missions to small bodies include some level of protection against contamination. As the committee has emphasized in this report, all missions to small bodies are currently Category I or Category II, thus imposing no cleanliness requirements. As stated in Finding 4, the committee endorses future reassessment in order to update the requirements and categorization for some small-bodies missions as necessary, as more is learned.

In briefing the committee, NASA's Planetary Protection Officer, J. Nick Benardini,[5] acknowledged that NASA is aware of commercial sector misperceptions about planetary protection policies and is making efforts to close knowledge gaps. Benardini described NASA plans to (1) update its planetary protection requirements, website, and mission planning tools; (2) establish a Planetary Protection Community of Practice; and (3) support and contribute to peer-reviewed planetary protection literature on planning and implementing planetary protection policies for missions. NASA also seeks to ensure that the commercial space industry understands that NASA's approach to planetary protection is a flexible, risk management approach to meeting planetary protection requirements. The committee supports NASA's efforts to clarify the commercial sector's understanding of NASA and COSPAR planetary protection policies.

Similarly, the committee encountered another problem identified repeatedly by previous reports on planetary protection—namely, whether and how planetary protection policies apply to commercial sector

---

[2] National Academies of Sciences, Engineering, and Medicine (NASEM), 2018, *Review and Assessment of Planetary Protection Policy Development Processes*, Washington, DC: The National Academies Press, https://doi.org/10.17226/25172, p. 88. (Hereinafter NASEM 2018 Report.)

[3] Planetary Protection Independent Review Board (PPIRB), 2019, *NASA Planetary Protection Independent Review Board (PPIRB): Report to NASA/SMD: Final Report*, Washington, DC: NASA, p. 4. (Hereinafter PPIRB Report.)

[4] NASEM, 2020, *Assessment of the Report of NASA's Planetary Protection Independent Review Board*, Washington, DC: The National Academies Press, https://doi.org/10.17226/26029. (Hereinafter NASEM 2020 Report.)

[5] J.N. Benardini and E. Seasly, 2021, "NASA Planetary Protection Status and Response to Previous CoPP Reports," Presentation to the Committee on Planetary Protection, November 30, Washington, DC: National Academies of Sciences, Engineering, and Medicine, https://www.nationalacademies.org/event/11-30-2021/docs/DF146D240CFD1BBB5517B2CD845E7343CE816A11CB18.

space activities in which NASA has no involvement.[6] In 2018, a committee of the National Academies of Sciences, Engineering, and Medicine (the National Academies) found that a "regulatory gap exists in U.S. federal law and poses a problem for U.S. compliance with the OST's obligations on planetary protection with regard to private sector enterprises."[7] It recommended that "Congress should address the regulatory gap by promulgating legislation that grants jurisdiction to an appropriate federal regulatory agency to authorize and supervise private-sector space activities that raise planetary protection issues."[8]

In 2019, NASA's Planetary Protection Independent Review Board noted a "lack of clarity concerning PP [planetary protection] requirements and implementation processes, particularly for non-NASA missions"[9] and recommended that "NASA should work with the Administration, the Congress, and the private sector space stakeholders to identify the appropriate U.S. Government agency to implement a PP regulatory framework."[10]

In 2020, a follow-up study by the National Academies' Committee to Review the Report of the NASA Planetary Protection Independent Review Board addressed the "regulatory gap" problem as follows:

> Problems persist with whether and how U.S. federal law regulates private-sector space activities for planetary protection purposes concerning launch, on-orbit, and re-entry activities. These problems create challenges for U.S. compliance with the Outer Space Treaty's obligations concerning the authorization and continual supervision of activities of non-governmental entities. The problems also undermine the private sector's need for a transparent and efficient legal and regulatory framework to support expanding private-sector exploration and uses of space.[11]

In response to that identified problem, the report goes on to provide the following recommendation:

> Recommendation: NASA should work with other agencies of the U.S. government, especially the Federal Aviation Administration, to produce a legal and regulatory guide for private-sector actors planning space activities that implicate planetary protection but that do not involve NASA participation. The guide should clearly identify where legal authority for making decisions about planetary protection issues resides, how the United States translates its obligations under the Outer Space Treaty into planetary protection requirements for non-governmental missions, what legal rules apply to private-sector actors planning missions with planetary protection issues, and what authoritative sources of information are available to private-sector actors that want more guidance on legal and regulatory questions.[12]

Also in 2020, the U.S. government issued its first-ever *National Strategy for Planetary Protection*, and the strategy stated that "the processes for approving and supervising private-sector space missions are currently unclear with regard to planetary protection."[13] Similarly, NASA and the White House Office of Science and Technology Policy convened a roundtable in 2021 to gather input from the commercial space

---

[6] For missions in which NASA is involved, NPR 8715.24 states that any partner, including both government agencies and private entities, to which NASA "may provide hardware, services, data, funding, deep-space communication, and other resources to non-NASA missions," should use "reasonable efforts to implement planetary protection measures generally consistent with the COSPAR Planetary Protection Policy and Guidelines or the planetary protection measures NASA would take for like missions."

[7] NASEM 2018 Report, p. 88.

[8] Ibid.

[9] PPIRB Report, p. 10.

[10] PPIRB Report, p. 18.

[11] NASEM 2020 Report, p. 6.

[12] Ibid.

[13] National Space Council, 2020, *National Strategy for Planetary Protection*, Washington, DC: Executive Office of the President, https://trumpwhitehouse.archives.gov/wp-content/uploads/2020/12/National-Strategy-for-Planetary-Protection.pdf, pp. 2–3.

industry on planetary protection. At this roundtable, private-sector representatives "sought clarification on which department or agency would enforce planetary protection requirements" and "suggested that a regulating agency may be necessary to enforce commercial mission requirements."[14]

The regulatory gap problem again arose in briefings to the committee by the NASA Planetary Protection Officer and a representative from the Federal Aviation Administration.[15] The longstanding and continuing confusion about the U.S. government's authority to apply and enforce planetary protection policies concerning commercial space activities prompted the committee to hold an open session with representatives from the executive and legislative branches of the U.S. government to discuss the problem.[16] For the committee, the session highlighted that the findings and recommendations in previous planetary protection reports on the regulatory gap problem have not been adopted.

> **Finding 8:** The application of planetary protection policies to private-sector space activities targeting small solar system bodies remains compromised by (1) misperceptions in the private sector about planetary protection requirements; and (2) confusion about the U.S. government's ability to apply and enforce planetary protection policies concerning nongovernmental space activities.

---

[14] Office of Science and Technology Policy/NASA Industry Roundtable for Planetary Protection Summary, 2021.

[15] Federal Aviation Administration and AST Commercial Space Transportation, 2022, "Go for Launch," presentation to the Committee on Planetary Protection, https://www.nationalacademies.org/event/01-19-2022/docs/D0D62386491A8B40C4AA5F7B4842F2A3ED2A338B821B.

[16] Roundtable Discussion on Regulatory Gap for Planetary Protection, Committee on Planetary Protection Meeting, Space Science Week 2022, March 23, 2022.

Appendixes

A

Statement of Task

The National Academies of Sciences, Engineering, and Medicine will appoint the Committee on Planetary Protection (CoPP) to operate as a long-term ad hoc committee. The disciplinary scope of CoPP includes the study of those aspects of planetary environments, the life sciences, spacecraft engineering and technology, and science policy relevant to the control of biological cross-contamination arising from the robotic spacecraft missions and the human exploration and utilization of solar system bodies.

CoPP will have two primary tasks:

1. To monitor progress in implementing the planetary protection guidelines associated with priority missions and programs identified in the planetary science decadal survey—*Vision and Voyages for Planetary Science in the Decade 2013–2022*—and in successor planetary science decadal surveys, and other relevant reports issued by the National Academies; and

2. To serve as a source of information and advice on those measures undertaken by robotic spacecraft and human exploration missions to protect the biological and environmental integrity of extraterrestrial bodies for future scientific studies and the means to preserve the integrity of Earth's biosphere when spacecraft return potentially hazardous extraterrestrial materials to Earth.

The committee will carry out its charge at its in-person and virtual meetings by gathering evidence from experts, deliberating, and, when necessary, by preparing short assessment reports detailing progress in areas relating to the National Aeronautics and Space Administration's (NASA's) planetary protection guidelines or new scientific and technical developments. Such reports may include findings and discussion of key activities undertaken by NASA as well as the status of its actions that relate to the state of implementation of priority missions and programs.

For other advisory activities that require a more in-depth review than is possible through the normal operation of the CoPP, the Space Studies Board (SSB), the Board on Life Sciences, the Aeronautics and Space Engineering Board, and NASA will negotiate a task for a separate ad hoc committee, taking advantage, as appropriate, of the expertise in the CoPP.

Through its regular meetings, the CoPP will also serve the secondary functions of:

1. Providing an independent, authoritative forum for the scientific community, the federal government, international space agencies, relevant private-sector entities and organizations, and the interested public to identify and discuss emerging issues in the scientific, technical, and engineering aspects of planetary protection policies and guidelines;

2. Identifying and prioritizing necessary research and development activities required to advance the development of planetary protection guidelines designed to ensure that the exploration and utilization of extraterrestrial environments is conducted responsibly; and

3. Providing a forum for interactions with the International Science Council's Committee on Space Research and other national and international organizations through the addition of international participants when appropriate and in coordination with the SSB.

The CoPP of the SSB shall conduct a study on planetary protection categorization of outbound-only missions to small bodies that addresses the following topics. In what follows, an "identifiable population" of solar system small bodies refers to a subset of solar system small bodies defined by ranges of measurable known parameters, such as (a) orbital elements, (b) spectroscopic classification, (c) activity, (d) composition, and/or (e) size. Objects yet to be discovered, whose properties fall into the defining ranges, are to be considered members of the corresponding identifiable population.

1. Are there identifiable populations of solar system small bodies that are sufficiently numerous, of sufficiently similar accessibility, and/or of sufficiently low relevance to the study of chemical evolution related to the search for extraterrestrial life that the contamination of *one* object in the population would reasonably be expected to have no tangible effect on the potential for scientific investigation using *other* objects in the population? If so, provide a list of these identifiable populations and their defining parameters;
2. For the populations identified in #1, is it appropriate to categorize all missions to objects in these as planetary protection Category I?
3. If, after the publication of the study, new information indicates that a previously defined identifiable population is sufficiently inhomogeneous with regard to planetary protection to warrant reassessment, what protocols should be followed in order to revise the defining parameter ranges and corresponding planetary protection categorizations?

The implications of the report findings will be consistent with the budget limitations provided by NASA at the time of study initiation. The study will gather input from stakeholders, including the planetary and astrobiology science communities, government agencies dealing with spaceflight and exploration, and the aerospace industry, including emerging commercial entities.

B

COSPAR Planetary Protection Requirements for Category I and Category II Missions

These categorizations are taken from COSPAR's Planetary Protection Policy.[1]

Category I includes any mission to a target body which is not of direct interest for understanding the process of chemical evolution or the origin of life. No protection of such bodies is warranted and no planetary protection requirements are imposed by this policy.

Category II missions comprise all types of missions to those target bodies where there is significant interest relative to the process of chemical evolution and the origin of life, but where there is only a remote chance that contamination carried by a spacecraft could compromise future investigations. The requirements are for simple documentation only. Preparation of a short planetary protection plan is required for these flight projects primarily to outline intended or potential impact targets, brief Pre- and Post-launch analyses detailing impact strategies, and a Post-encounter and End-of-Mission Report which will provide the location of impact if such an event occurs. Solar system bodies considered to be classified as Category II are listed in the Appendix to this document.

Category-specific listing of target body/mission types:

Category I: Flyby, Orbiter, Lander: Undifferentiated, metamorphosed asteroids; Io; others to be defined (TBD)

Category II: Flyby, Orbiter, Lander: Venus; Moon (with organic inventory); Comets; Carbonaceous Chondrite Asteroids; Jupiter; Saturn; Uranus; Neptune; Ganymede*; Callisto; Titan*; Triton*; Pluto/Charon*; Ceres; Kuiper Belt objects > half the size of Pluto*; Kuiper Belt objects < half the size of Pluto; others TBD

* The mission-specific assignment of these bodies to Category II must be supported by an analysis of the "remote" potential for contamination of the liquid-water environments that may exist beneath their surfaces (a probability of introducing a single viable terrestrial organism of $< 1 \times 10^{-4}$), addressing both the existence of such environments and the prospects of accessing them.

---

[1] "COSPAR Policy on Planetary Protection," prepared by the COSPAR Panel on Planetary Protection and approved by the COSPAR Bureau on June 3, 2021, https://cosparhq.cnes.fr/assets/uploads/2021/07/PPPolicy_2021_3-June.pdf.

C

Acronyms and Abbreviations

AU	astronomical unit
C-G	Churyumov-Gerasimenko
CoPP	Committee on Planetary Protection
COSPAR	Committee on Space Research
D/H	deuterium-to-hydrogen ratio
EPOXI	Extrasolar Planet Observation and Deep Impact Extended Investigation
ESA	European Space Agency
FAA	Federal Aviation Administration
GCR	galactic cosmic radiation
IRTF	Infrared Telescope Facility
ISM	interstellar medium
JFC	Jupiter family comet
JPL	Jet Propulsion Laboratory
KBO	Kuiper Belt object
MBA	Main Belt asteroid
MBC	Main Belt comet
MIRSI	Mid-InfraRed Spectrometer and Imager
NASA	National Aeronautics and Space Administration
NEA	near-Earth asteroid, synonymous with NEO
NEAR	Near Earth Asteroid Rendezvous
NEO	near-Earth object
NIR	near-infrared
NPR	NASA Procedural Requirement
OSIRIS-REx	Origins, Spectral Interpretation, Resource Identification, Security, Regolith Explorer
PDS	Planetary Data System
PPIRB	Planetary Protection Independent Review Board

SSB	Space Studies Board
STScI	Space Telescope Science Institute
SwRI	Southwest Research Institute
TBD	to be defined
TNO	Trans-Neptunian object
UVC	ultraviolet C

D

Committee and Staff Biographies

JOSEPH K. ALEXANDER, *Co-Chair*, is a consultant in science and technology policy. He was a senior program officer with the National Academies of Sciences, Engineering, and Medicine's Space Studies Board (SSB) from 2005 until 2013, and he served as SSB director from 1998 until November 2005. Prior to joining the National Academies, he was deputy assistant administrator for science in the Environmental Protection Agency's Office of Research and Development, where he coordinated a broad spectrum of environmental science and led strategic planning. From 1993 to 1994, he was associate director of space sciences at the National Aeronautics and Space Administration (NASA) Goddard Space Flight Center (GSFC) and served concurrently as acting chief of the Laboratory for Extraterrestrial Physics. From 1987 until 1993, he was assistant associate administrator at NASA's Office of Space Science and Applications, where he coordinated planning and provided oversight of all scientific research programs. He also served from 1992 to 1993 as acting director of life sciences. Prior positions include deputy NASA chief scientist, senior policy analyst at the White House Office of Science and Technology Policy, and research scientist at GSFC. His research interests were in radio astronomy and space physics. He has a BA and an MA in physics from the College of William and Mary. His book, *Science Advice to NASA: Conflict, Consensus, Partnership, Leadership*, was published in 2017. He has served on multiple committees of the National Academies, including the Committee to Review the Report of the NASA Planetary Protection Independent Review Board (chair), the Committee on the Review of Planetary Protection Policy Development Processes (chair), and the Committee on the Review of NASA's Planetary Science Division's Restructured Research and Analysis Program (member).

AMANDA R. HENDRIX, *Co-Chair*, is a senior scientist with the Planetary Science Institute. Her research interests focus on moons and small bodies in the solar system to understand composition, activity, and evolution. Hendrix is director of NASA's SSERVI TREX node, previously a co-investigator on the Cassini UVIS and Lunar Reconnaissance Orbiter LAMP teams, was a co-investigator on the Galileo UVS team and served as the Cassini deputy project scientist. In 2016, she published a book (Penguin/Random House) with co-author Charles Wohlforth, *Beyond Earth: Our Path to a New Home in the Planets*, a discussion of the technological, medical, and social hurdles to overcome in considering a human space establishment in the outer solar system. Hendrix is co-chair of the Roadmaps to Ocean Worlds group, serves as a steering committee member of the Outer Planets Assessment Group, and was a member of the Planetary Protection Independent Review Board. She earned her PhD in aerospace engineering with an emphasis in planetary science from the University of Colorado. Hendrix has served on various National Academies' committees, including the Committee on the Review of Progress Toward Implementing the Decadal Survey Vision and Voyages for Planetary Sciences.

ANGEL ABBUD-MADRID is director of the Center for Space Resources at the Colorado School of Mines, where he leads a research program focused on the human and robotic exploration of space and the utilization of its resources. He is also director of the Space Resources Graduate Program aimed at educating scientists, engineers, economists, entrepreneurs, and policy makers in the field of extraterrestrial resources. Abbud-Madrid has more than 30 years of experience conducting experiments in NASA's low-gravity facilities, such as drop towers, parabolic-flight aircraft, the space shuttle, and the

International Space Station and received the NASA Astronauts' Personal Achievement Award for his contributions to the success of human space flight. He is currently president of the Space Resources Roundtable, an international organization focused on lunar, asteroidal, and planetary resources studies. In addition, Abbud-Madrid is an observer and technical panel member of The Hague International Space Resources Governance Working Group. He received his PhD in mechanical and aerospace engineering from the University of Colorado Boulder.

ANTHONY COLAPRETE is a planetary scientist at the NASA Ames Research Center in the Space Sciences Division. His research interests include planetary exploration, in situ resource utilizations, volatiles, and radiative transfer. With more than 20 years of experience, he has worked on a variety of space projects ranging from sounding rockets and space shuttle flights to micro and small satellites. Prior to joining NASA Ames, he was a principal investigator at the SETI Institute, a National Research Council associate at NASA Ames, and a space scientist at the Laboratory for Atmospheric and Space Physics and the Colorado Space Grant Consortium. Colaprete is the recipient of the 2016 H. Julian Allen Award. He received his PhD in astrophysical, planetary, and atmospheric science from the University of Colorado.

MICHAEL J. DALY is a professor in the Department of Pathology at the Uniformed Services University of the Health Sciences School of Medicine in Bethesda, Maryland. He is an expert in the study of bacteria belonging to the family *Deinococcaceae*, which are some of the most radiation-resistant organisms yet discovered. He received his PhD in molecular biology at Queen Mary University of London. He has served on multiple National Academies' committees, including the Committee on Planetary Protection Requirements for Sample Return Missions from Martian Moons, the Committee on Planetary Protection Standards for Icy Bodies in the Outer Solar System, the Committee on Planetary Protection Requirements for Venus Missions, the Committee on the Origins and Evolution of Life, the Committee on the Astrophysical Context of Life, and the Committee for the Task Group on the Forward Contamination of Europa.

DAVID P. FIDLER is a senior fellow for cybersecurity and global health at the Council on Foreign Relations and the James Louis Calamaras Professor of Law (emeritus) at the Indiana University Maurer School of Law. He works on international law and global governance issues across many policy areas, including cyberspace, global health, outer space, national security, environmental protection, terrorism, and weapons of mass destruction. Current activities include research on the COVID-19 pandemic, cybersecurity law, and emerging challenges in global space governance. He is the recipient of a Fulbright New Century Scholar Award. Fidler received his JD from Harvard Law School. He has served on numerous National Academies' committees, including the Committee to Review the Report of the NASA Planetary Protection Independent Review Board and the Committee to Review the Planetary Protection Policy Development Processes.

SARAH A. GAVIT is project manager for the VenSAR Project at the Jet Propulsion Laboratory (JPL). She has more than 35 years of engineering and management experience. Previous assignments at JPL include serving as deputy division manager for the Communications, Radar and Tracking Division, assistant director for engineering and science, project manager for the Dawn mission and the Deep Space 2 Mars Microprobe Project, project system engineer for the Prometheus and Kepler missions, fault protection system engineer for the Cassini mission, and as the Mars System Sterilization Study lead. Early in her career at Martin Marietta, Gavit was a mission and system engineer for the Magellan mission to Venus. Gavit operated her own business as a private consultant to NASA for spacecraft system engineering and project management, and frequently served on technical, management, and cost panels for space mission evaluations. Gavit received her MS in aeronautical and astronautical engineering from the Massachusetts Institute of Technology.

ANDREW D. HORCHLER is the principal research scientist at Astrobotic, where he leads the research and development of robotics hardware and software for advanced space applications. Horchler has fielded more than a dozen mobile robot platforms over the past 20 years and has published more than 60 papers, proceedings, and patents. His robots have been tested on simulated lunar regolith at NASA Glenn Research Center's Simulated Lunar Operations laboratory, on tortuous rubble piles and desert terrain for NASA and National Institute of Standards and Technology field tests, and have flown in caves and icy lava tubes. He leads the development of a navigation sensor for precision landing that will fly on Astrobotic's first lunar mission and a hazard detection sensor that will safely land NASA's VIPER rover on the South Pole of the Moon in 2023. Horchler also supports rover system development and served as Principal Investigator for Astrobotic's lunar "CubeRover" platform as well as software to aid mission planners in formulating rover missions under the unique lighting conditions at the poles of the Moon. Prior to joining Astrobotic, he was the technical lead for Case Western Reserve's Defense Advanced Research Project Agency Urban Challenge self-driving car team where he led the creation and testing of driving behaviors and developed real-time trajectory planning and mapping algorithms.

EUGENE H. LEVY is the Andrew Hays Buchanan Professor of Astrophysics in the Department of Physics and Astronomy at Rice University. His research interests focus on theoretical cosmic physics, with emphasis on elucidating mechanisms and processes that underlie physical phenomena in planetary and astrophysical systems. Levy's research also includes the generation and influences of magnetic fields in natural bodies, including Earth, the Sun, and planets, the theory of cosmic rays, and the theory of physical processes associated with the formation of the solar system, stars, and other planetary systems. Prior to joining Rice University, he served in various capacities at the University of Arizona, including dean of the College of Science, head of the Planetary Science Department, director of the Lunar and Planetary Laboratory, and professor of planetary science. Levy has won multiple awards, including the Alexander von Humboldt-Stiftung Senior Scientist Award by the Federal Republic of Germany, the Martin Luther King, Jr. Distinguished Leadership Award through the University of Arizona, and the NASA Distinguished Public Service Medal. He received his PhD in physics from the University of Chicago. Levy has served on various committees at the National Academies, including the Committee on the Review of Planetary Protection Policy Development Processes, the Committee for US-USSR Workshop on Planetary Sciences, the ad hoc Panel on Mars Sample Return, and the Planetary and Lunar Exploration Task Group.

ROBERT E. LINDBERG, JR., is an independent consultant with more than 35 years of experience as an accomplished aerospace executive and entrepreneur that spans government, aerospace industry, start-ups, academic, and nonprofit sectors. Lindberg's background and experience includes spacecraft and launch vehicle design; entry, descent, and landing; and planetary protection. Prior to his current position, he served as vice president of two small space companies: Moon Express and Vector Launch. From 2003 to 2012, he was president and executive director of the National Institute of Aerospace (NIA). Prior to co-founding NIA, Lindberg was senior vice president with Orbital Sciences Corporation (now a division of Northrop Grumman). Lindberg was a former member of the NASA Advisory Council Science Committee and chaired its Planetary Protection Subcommittee. He is affiliated with the American Institute of Aeronautics and Astronautics (fellow) and the American Astronautical Society (fellow and past president). Lindberg received numerous honors including the Egleston Medal from Columbia University and the Engineering Achievement Award from the University of Virginia. He has served on committees and panels for NASA, the Naval Studies Board of the National Academies, the National Security Space Architect, the Federal Aviation Administration, and the International Council on Science's Committee on Space Research. Lindberg received his EngScD in mechanical engineering from Columbia University. He served on the National Academies' Committee on the Navy's Needs in Space for Providing Future Capabilities.

MARGARITA M. MARINOVA[1] works for Amazon's Project Kuiper. She has worked on improving rocket capabilities and reusability, gaining deeper understanding of Earth and its planetary neighbors, and applying these advancements to improve life on Earth. Marinova has worked at Airbus Space Propulsion in engine nozzle research and development and at NASA Ames Research Center as a planetary scientist and has studied a diverse variety of extreme environments, including the High Arctic, the Sahara Desert in Egypt, and the Dry Valleys of Antarctica. Most recently she was at SpaceX as a propulsion systems responsible engineer for the vertical takeoff and landing F9R-Dev vehicle, vehicle responsible engineer for internal research and the reusability program, and senior Mars development engineer working on mission architecture and vehicle design for the Starship vehicle and its planetary missions. Marinova received her PhD in planetary science from the California Institute of Technology. She served on the National Academies' Committee to Review the NASA's Planetary Protection Independent Review Report.

A. DEANNE ROGERS is an associate professor with the Department of Geosciences at Stony Brook University and editor of the *Journal of Geophysical Research: Planets*. Prior to joining Stony Brook, she was a postdoctoral fellow at the California Institute of Technology. Her research interests include using remote sensing techniques, statistical methods, laboratory spectroscopy, and thermal modeling to investigate a wide range of planetary surface processes. She manages the Earth and Planetary Remote Sensing Laboratory under the Stony Brook Center for Planetary Exploration. Rogers is the recipient of numerous awards, including the NASA Planetary Science Division Early Career Fellow, the NASA Group Achievement Award for Mars Exploration Rovers, and the NASA Group Achievement Award for the 2001 Odyssey THEMIS. She received her PhD in geological sciences from Arizona State University.

GERHARD H. SCHWEHM has more than 30 years of experience working for the European Space Agency (ESA) (retired) in various positions. This includes serving as Rosetta Mission Manager from 2004 to 2013, head of the Solar System Science Operations Division at ESA-European Space Astronomy Centre from 2007 to 2011, and head of the Planetary Missions Division at ESA-European Space Research and Technology Centre from 2001 to 2007. During his time at ESA, Schwehm served as a member of the Interagency Space Debris Working Group, ESA representative for the NASA Planetary Protection sub-group, and a member of the ESA Planetary Protection Working Group. He is an ex-officio of numerous mission and payload reviews and selection panels for ESA, NASA, and DLR. Schwehm received his PhD in applied physics from the Ruhr-Universitat Bochum.

TRISTA J. VICK-MAJORS is an assistant professor in the Department of Biological Sciences at Michigan Technological University and a member of the SALSA (Subglacial Antarctic Lakes Scientific Access) Science Team. She currently serves on the science advisory board for the United States Ice Drilling Program. Prior to joining Michigan Technological University, she was a postdoctoral research scientist at l'Université du Québec à Montréal and at the University of Montana's Flathead Lake Biological Station. Her main research interests focus on microbial life and biogeochemical processes in and under ice, microbial growth under oligotrophic and energy-limited conditions in aquatic systems, and clean access to pristine subglacial aquatic environments. She has participated in three research expeditions to study permanently ice-covered lakes in the Antarctic McMurdo Dry Valleys, including the only study of the region during the onset of the austral winter, and three that accessed subglacial water under ~1 km of ice on the West Antarctic Ice Sheet and Ross Ice Shelf as part of the SALSA and WISSARD (Whillans Ice Stream Subglacial Access Research and Drilling) projects. She earned her PhD in ecology and environmental sciences from Montana State University. Vick-Majors served on the National Academies' Committee for the Review of the NASA Independent Review Board and participated in a workshop of experts convened by the Division on Earth and Life Studies on Understanding and Responding to Global Health Security Risks from Microbial Threats in the Arctic.

---

[1] Recused from this study.

STAFF

DANIEL NAGASAWA, *Study Director*, joined the SSB in 2019 and is a program officer. Before joining the SSB, he was a graduate research assistant specializing in stellar astrophysics, measuring the abundance of elements in the atmospheres of very old, metal-poor stars. Nagasawa began his research career as an undergraduate research assistant for the Cryogenic Dark Matter Search. When he began graduate school, he transitioned to designing and evaluating astronomical instrumentation, specifically ground-based spectrographs. He went on to specialize in high-resolution stellar spectroscopy and applied these techniques on stars in ultra-faint dwarf satellite galaxies of the Milky Way to study the chemical history of the galaxy as part of the Dark Energy Survey (DES). He also developed skills in education and public outreach by teaching an observational astronomy course and writing for an outreach initiative for DES. Nagasawa earned his PhD in astronomy and his MS in physics at Texas A&M University; he earned his BS in physics with a concentration in astrophysics from Stanford University.

NANCY CONNELL is a senior scientist in the Board on Life Sciences at the National Academies. Trained in microbial genetics at Harvard, Connell's work has focused on advances in life sciences and technology and their application to a number of developments in the areas of biosecurity, biosafety, and biodefense. She has had a long-standing interest in the development of regulatory policies associated with biocontainment work and dual-use research of concern. Connell is a past member of the Board on Life Sciences and the Committee on International Security and Arms Control, and a National Associate of the National Academies, where she has served on more than 15 National Academies' committees. Among other national and international committees, she served on the U.S. National Science Advisory Board for Biosecurity. Connell has considerable experience and interest in pedagogy, with an international focus on ethics education and the responsible conduct of research. She was a senior scientist at the Johns Hopkins Center for Health Security and a professor in the Department of Environmental Health and Engineering at the Johns Hopkins Bloomberg School of Public Health from 2018 to 2021. From 1992 to 2018, Connell was an investigator in microbial genetics and drug discovery at Rutgers New Jersey Medical School, finishing her long career there as a professor in the Division of Infectious Disease and director of research in the Department of Medicine.

ALEXANDER BELLES is a 2022 Christine Mirzayan Science & Technology Policy Graduate Fellow with the SSB. He is a PhD candidate in the Department of Astronomy and Astrophysics at the Pennsylvania State University. His graduate work has focused on panchromatic studies of nearby galaxies and the wavelength dependent effects of interstellar dust. During his graduate career, Belles has been a member of the Science Operations Team for the Neil Gehrels Swift Observatory, a NASA space-based observatory with three telescopes used to study gamma-ray bursts. As an undergraduate, he started doing research by studying lithium depletion in open star clusters. Previously, Belles received his BA in physics and mathematics from the State University of New York College at Geneseo.

MEGAN CHAMBERLAIN joined the SSB and the Aeronautics and Space Engineering Board (ASEB) as a senior program assistant in 2019. Chamberlain began her career at the National Academies in 2007 working for the Transportation Research Board in the Cooperative Research Programs. She has assisted with meeting facilitation and administrative support of hundreds of research projects over the course of her career. Chamberlain attended the University of the District of Columbia and majored in psychology.

COLLEEN N. HARTMAN joined the National Academies in 2018 as director for both the SSB and the ASEB. After beginning her government career as a presidential management intern under Ronald Reagan, Hartman worked on Capitol Hill for House Science and Technology Committee Chairman Don Fuqua, as a senior engineer building spacecraft at NASA GSFC, and as a senior policy analyst at the White House. She has served as Planetary Division director, deputy associate administrator, and acting associate administrator at NASA's Science Mission Directorate, as deputy assistant administrator at the National

Oceanic and Atmospheric Administration, and as deputy center director and director of science and exploration at NASA GSFC. Hartman has built and launched scientific balloon payloads, overseen the development of hardware for a variety of Earth-observing spacecraft, and served as NASA program manager for dozens of missions, the most successful of which was the Cosmic Background Explorer (COBE). Data from the COBE spacecraft gained two NASA-sponsored scientists the Nobel Prize in physics in 2006. She also played a pivotal role in developing innovative approaches to powering space probes destined for the solar system's farthest reaches. While at NASA Headquarters, she spearheaded the selection process for the New Horizons probe to Pluto. She helped gain administration and congressional approval for an entirely new class of funded missions that are competitively selected, called "New Frontiers," to explore the planets, asteroids, and comets in the solar system. She has several master's degrees and a PhD in physics. Hartman has received numerous awards, including two prestigious Presidential Rank Awards.